BACTERIOLOGY RESEARCH DEVELOPMENTS

Pseudomonas aeruginosa

A REVIEW AND DIRECTIONS FOR RESEARCH

BACTERIOLOGY RESEARCH DEVELOPMENTS

Additional books and e-books in this series can be found on Nova's website under the Series tab.

BACTERIOLOGY RESEARCH DEVELOPMENTS

Pseudomonas aeruginosa

A REVIEW AND DIRECTIONS FOR RESEARCH

CHEE FAH WONG
AND
HAMIDAH IDRIS
EDITORS

nova science publishers
New York

Library of Congress Cataloging-in-Publication Data

ISBN: 978-1-53615-100-7
Library of Congress Control Number: 2019947958

Published by Nova Science Publishers, Inc. † New York

CONTENTS

PREFACE

Classified as an opportunistic pathogen, *Pseudomonas aeruginosa* live widely in nature and regarded as one of the oldest genus that have been discovered. Reports on *P. aeruginosa* infection have been published widely and it is still one of the major threats in medical field. Apart from that, the ability of this microorganism to produce useful substances has drawn attention from scientists to look into its potentials in medical, industrial and environmental applications. Nevertheless, research work in this field is growing actively and more discovery has yet to be uncovered.

Chapter 1 discusses recent advancements in application of *P. aeruginosa* proteins by employing strategies to obtain overexpression of both industrially- and medically-important proteins. In this chapter, the nuisance of *P. aeruginosa* gene expression is highlighted and the latter issue is succeeded by introduction of several genetic elements and construction of *Escherichia-Pseudomonas* shuttle expression vectors to achieve expression of active and soluble proteins in desirable amount.

Chapter 2 covers the capability of *P. aeruginosa* to form biofilms. Compared to free living cells, biofilm formation allows *P. aeruginosa* to be more resistant against antibacterial agents and more powerful in attacking host cells. Biofilm development is described in a comprehensive way from physico-chemical, biological and molecular perspectives.

Surface interactions and antimicrobial resistance are among the focus in this chapter.

Quorum sensing is an intracellular communication system which allow bacteria to response to the changes in environment. Chapter 3 reviewed quorum sensing in *P. aeruginosa* and the understanding of known systems in this topic is outlined. Due to the unique interactions and effect of inhibitors, QS system can be used for future application such as in medical field for alternative therapeutic intervention against *P. aeruginosa.*

Heavy metal pollution caused by industries is affecting the environment and the effective ways has to be taken to solve this problem. Research had proven that *P. aeruginosa* can play role in bioremediation due to the ability to reduce heavy metals content through the process of biosorption. In Chapter 4, mechanisms of biosorption process by *P. aeruginosa* is outlined in detail and the latest research update, challenges and future perspectives is also discussed.

Chapter 5 begins by laying out the skin ulcers infection by *P. aeruginosa*, and looks on the effectiveness of photodynamic therapy in solving the problem. Research data from both *in vitro* and *in vivo* experiments is highlighted and showed that photodynamic therapy could serve as an alternative to antimicrobial therapy.

Finally, Chapter 6 discusses about lipoxygenase, the candidate for novel virulence factor of *P. aeruginosa* that has been identified by several studies. Description on the characteristics and catalytic properties and is well explained and several hypotheses pertaining biological relevance of the enzyme are extensively discussed.

We would like to express our heartfelt appreciation to all authors of the chapters in this book for their effort, commitment and cooperation in making this book a reality. May this book reach and bring benefit to readers and contribute to advancement in *Pseudomonas* research area.

Chee Fah Wong
Hamidah Idris

In: *Pseudomonas aeruginosa* ISBN: 978-1-53616-100-7
Editors: C. Fah Wong and H. Idris

Chapter 1

DEVELOPMENT OF GENETIC TOOLS FOR PROTEIN OVEREXPRESSION FROM *Pseudomonas*

Chee Fah Wong
Department of Biology, Faculty of Science and Mathematics,
Universiti Pendidikan Sultan Idris, Tanjung Malim, Perak, Malaysia

ABSTRACT

Overexpression of a number of industrially- and medically-important proteins is always a challenge for most researchers as genes from bacterial hosts especially from the opportunistic pathogen, *Pseudomonas aeruginosa*, as well as from other Pseudomonads were reported to be poorly expressed in *E. coli*. Examples continue to be published demonstrating that certain *Pseudomonas* proteins cannot be expressed at satisfactory levels in *E. coli* owing to their promoter variations, absence of specific molecular activators and poor protein translation. Nevertheless, the extension of *E. coli*-based technology to *Pseudomonas* can be used to overcome such barriers. In this chapter, the evolution and modification of genetic tools ranging from types of promoters, elements for interspecies transfer, origins of replications, regulatory genes, genes facilitating in genome integration and host strains, used in the construction of *Escherichia-Pseudomonas* shuttle expression vectors will be explored and discussed in detail. Protein overexpression data as documented by the constructed shuttle expression vectors in this chapter, may offer a new strategy to express abundant amount of active proteins which can be an important determinant factor for successful nuclear magnetic resonance (NMR) and X-ray crystallography studies especially involving proteins from *Pseudomonas* and its related genus.

INTRODUCTION

The *Pseudomonas* species and closely related bacteria include members that are of considerable importance in agriculture, industry and medicine. Many aspects of their biology, biochemistry and pathogenicity have therefore been reasonably well worked out. Many of these advances would not have been possible without the development of a myriad of genetic and molecular tools, all of which were derived in individual laboratories as none of them is commercially available. These tools include broad-host-range cloning and expression vectors, vehicles to deliver genes into the bacterial chromosome, and techniques to monitor gene expression in diverse environments.

The majority of expression vectors that are used on a regular basis to express genes in *Pseudomonas* species are based or RSF1010, RK2 or pRO1600 (Figure 1.1) and over the past few years most newly described vectors have in fact been modifications of previously published constructs (Schweizer, 2001). Examples continue to be published demonstrating that certain *Pseudomonas* proteins cannot be expressed at satisfactory levels in *E. coli,* but that the extension of *E. coli*-based technology to *Pseudomonas* can be used to overcome such barriers. Additionally, genes from bacterial hosts such as the opportunistic pathogen, *P. aeruginosa*, as well as from other Pseudomonads were reported to be poorly expressed in *E. coli* (Minton et al., 1983). Few possible reasons were known to be:

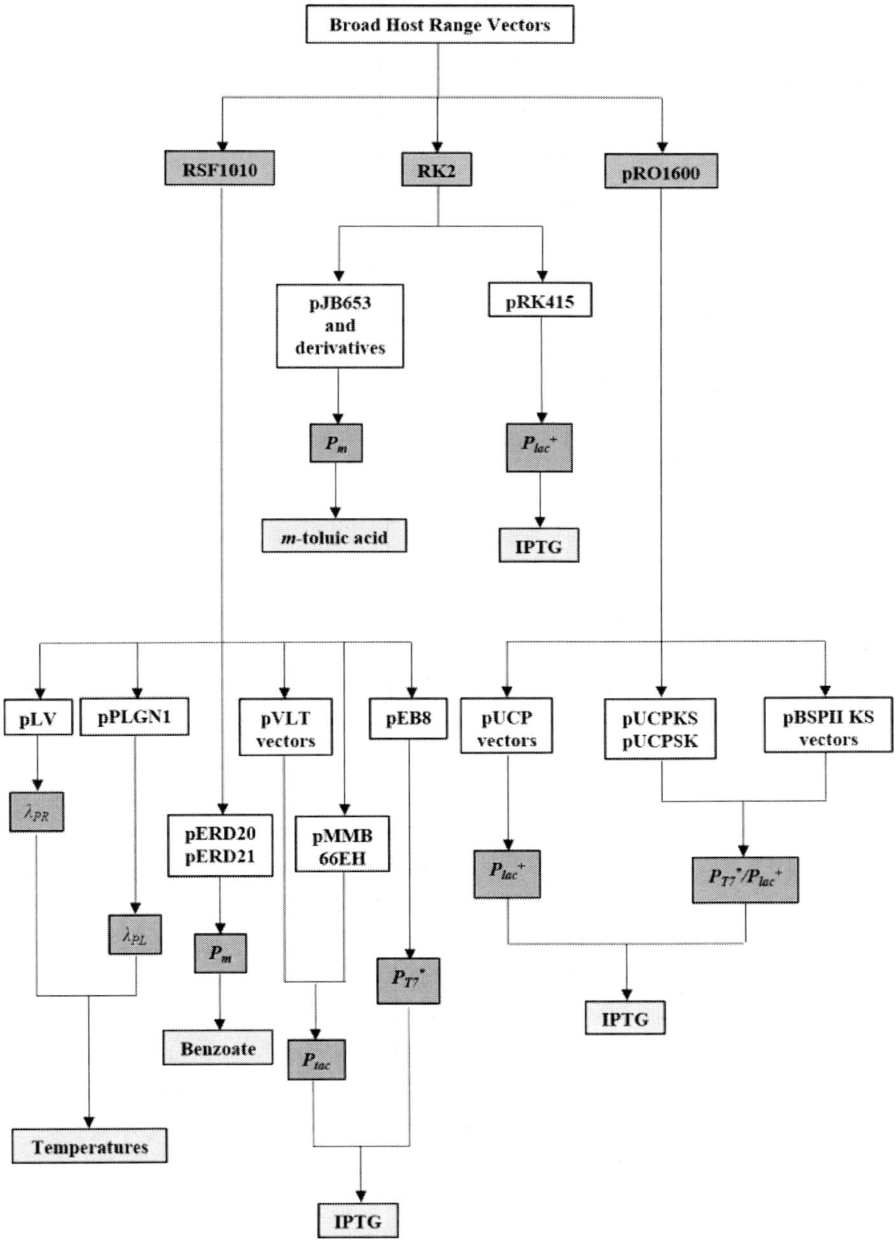

Figure 1.1. Selected broad-host-range expression vectors. Replicons, promoters and inducers are represented by color-coded boxes ▢, ▢ and ▢, respectively.

- Significant variation of the *Pseudomonas* promoter structure from the *E. coli* consensus sequence (Rothmel et al., 1991).
- Most of the genes that have been investigated are positively regulated and require specific activator molecules that are presumably absent in *E. coli* (Pasloske et al., 1989).
- Some target proteins are produced in small quantities even in the presence of large quantities of specific and stable mRNA. This poor translation is most likely due to the particularly high G+C content of *Pseudomonas* DNA (67% for *P. aeruginosa)* which dictates a unique codon usage and results in an unfavorable distribution of rare codons (West and Iglewski, 1988).

Although there are many different broad-host-range vectors available, the majority of them are based on the RSF1010 (Scholz et al., 1989), RK2 (Thomas, 1981) or pRO1600 (West et al., 1994) replicons (Figure 1.1). To achieve regulated expression of cloned genes, many of these vectors contain promoters that can be regulated, such as the T7 promoter from the T7 bacteriophage gene 10, the leftward (λPL) and rightward (λPR) promoters from *Escherichia coli* phage λ, P_m and P*u* from the *Pseudomonas putida* TOL plasmid pWWO, and the *E. coli lac* operon-based promoters P_{lac}, P_{tac} and P_{trc} (Warren et al., 2000). Of these, the T7 and *lac* operon-based promoters have traditionally found the most widespread use, because the biology of these two systems is adaptable to many different bacteria and the required materials are readily available in most laboratories.

P_{LAC}-Based Expression System

Although the expression of cloned genes was regulated fairly well in most expression constructs, numerous researchers have found that significant amounts of protein can be produced even under non-inducing and repressed conditions when using many of these vectors. Some of these problems stem from the rudimentary regulatory regions present on expression vectors (Schweizer, 2001). For example, most of the *lac*-based expression vectors contain only one of the *lac* operators required for efficient Lac-repressor-mediated repression of transcription. This dilemma was resolved by Warren et al. (2000) with inclusion of an auxiliary operator, resulted in increased or almost complete repression. This new expression vector, pLAC11, was designed to be more regulable and thus more tightly repressible when grown under repressed conditions. The tight regulation of pLAC11 was achieved by utilizing the O3 auxiliary operator, CAP binding site, promoter, and O1 operator that occur in the wild-type *lac* control region.

The pLAC11 vector can be used to conduct physiologically relevant studies in which the cloned gene is expressed at levels comparable to that obtainable from the chromosomal copy of the gene in question. In experiments in which a bacterial cell contained both a null allele in the chromosome and a second copy of the wild-type allele on pLAC11, Warren et al. (2000) observed that cells grown under repressed conditions exhibited the null phenotype while cells grown under induced conditions exhibited the wild-type phenotype. Two multipurpose derivatives of pLAC11, pLAC22, and pLAC33 have also been constructed to fulfil different experimental needs. These newly described *E. coli* expression vectors allow extremely tight regulation by the Lac repressor and could easily be modified for use in *Pseudomonas*.

[*]Regulated expression from the P_{T7} promoter requires a specialized host strain containing a plasmid or chromosomal copy of T7 polymerase.

[+]Regulated expression from P_{lac} is only possible in host strains that contain a chromosomal copy of *lacI*. IPTG, isopropyl-β-D thiogalacto-pyranoside.

P_{T7}-Based Expression System

Alternatively, a gene expression system based on bacteriophage T7 RNA polymerase has also been developed by Brunschwig and Darzins (1992) as well as Hoang et al. (2000) to combat the level of basal transcription or leakiness that occurred in the absence of induction in *Pseudomonas* sp. and *E. coli*. T7 RNA polymerase is highly selective for its own promoters, which do not occur naturally in *Escherichia coli*. A relatively small amount of T7 RNA polymerase provided from a cloned copy of T7 gene 1 is sufficient to direct high-level transcription from a T7 promoter in a multicopy plasmid. Such transcription can proceed several times around the plasmid without terminating, and can be so active that transcription by *E. coli* RNA polymerase is greatly decreased.

In earlier experiments, Studier and Moffatt (1986) have used two ways to deliver active T7 RNA polymerase to the cell; infection by a lambda derivative that carries gene 1, or induction of a chromosomal copy of gene 1 under control of the *lac*UV5 promoter. When gene 1 is delivered by infection, very toxic target genes can be maintained silent in the cell until T7 RNA polymerase is introduced, when they rapidly become expressed at high levels. When gene 1 is resident in the chromosome, even the very low basal levels of T7 RNA polymerase present in the uninduced cell can prevent the establishment of plasmids carrying toxic target genes, or make the plasmid unstable. But if the target plasmid can be maintained, induction of chromosomal gene 1 can be a convenient way to produce large amounts of target RNA and/or protein. T7 RNA polymerase seems to be capable of transcribing almost any DNA linked to a T7 promoter, so the T7 expression system should be capable of transcribing almost any gene or its complement in *E. coli*.

PROTEIN EXPRESSION IN *Pseudomonas*

Gram-negative rods *Pseudomonas* sp. are known for their rapid growth and ability to secrete various industrially-important proteins (Joshi et al., 2014). *Pseudomonas* sp. typically *P. fluorescens*, *P. aeruginosa* and *P. putida* were shown to be good alternatives to *E. coli* in recombinant protein expression since the formers are able to achieve a yield comparable or higher than *E. coli* systems (Loeschcke and Thies, 2015). *Pseudomonas* sp. can be transformed with plasmid DNA by chemical methods, electroporation or by plasmid conjugation, as conducted in *E. coli* (Huang and Wilks, 2017). Additionally, *Pseudomonas* contains relatively high G+C content that makes the latter an ideal host for expression of certain restriction endonucleases (Huang and Wilks, 2017).

In recent years, *Pseudomonas fluorescens* was used to produce remarkably high yield of protein (Liu et al., 2017). Retallack and collegues (2012) developed a proprietary *P. fluorescens* expression system for recombinant protein expression. The *P. fluorescens* system exhibited advantages in protein expression due to its ability for high production of recombinant proteins and possesses the ability to achieve high cell density similar to *E. coli*. Besides, the protein production of *P. fluorescens* is less dependent on oxygen concentration and lacks accumulation of acetate during the expression (Zhang et al., 2018). Consequently, a strict control of aeration and glucose concentration are not required during fermentation of *P. fluorescens*. Several examples of successes have been reported for protein expression in *P. fluorescens* namely the expression of nitrilase by Chen et al. (2015), identification of signal peptides in *P. fluorescens* through utilization of transposon mutant library and *E. coli* PhoA (Retallack et al., 2007) and mechanistic elucidation of soluble protein expression in periplasmic space (Wilton et al., 2018).

P. aeruginosa is yet another common host for recombinant protein expression due to its native and well-studied secretion systems especially for type II and type III (Tosi et al, 2014). Vareechon et al. (2017) constructed a novel expression system in order to study the extracellular production of *P. aeruginosa via* and its type III. Type III substrate exotoxin S (ExoS) transitioned to the target protein and the inducible promoter, P_{tac} controlled the expression of a common regulator of *P. aeruginosa*. As a result, activation of the regulator was allowed by plasmid and the target protein was shown to be secreted under calcium deprivation condition generated by the addition of EGTA.

P. putida also been studied as a host for recombinant protein expression. *P. putida* KT2440 was used in the development of genetic manipulation tools that are useful for engineered recombinant protein expression analysis (Cook et al., 2018; Dammeyer, 2011). This strain was used to study the efficiency of the production of soluble recombinant single chain Fv fragment (scFv). An expression vector was constructed based on RK2 broad host range plasmid with either P_{tac} or P_m promoter which worked very well in the host. The expression vector possesses G10L ribosome binding site with epsilon enhancer and a leader sequence to direct the expressed recombinant protein to its periplasm. Perisplasmic expression is essentially necessary for proteins with disulphide bonds as the oxidative condition present in periplasmic space is more conducive for the proper formation of disulphide bonds. The scFv, which developed as the expression model, was also purified in single step purification method (Dammeyer et al., 2011).

INTEGRATION-PROFICIENT VECTORS, MINI-CTX1, MINI-CTX2 AND MINI-CTX-T7

The advanced stage of vector construction for protein expression in Pseudomonas was reported by Hoang et al. (2000) who developed improved vectors for integration of exogenous DNA fragments at a defined site within the genome of *Pseudomonas aeruginosa*. These vectors contained:

- a tetracycline (*tet*) selectable marker
- an *oriT* for conjugation-mediated plasmid transfer
- the pMB1-derived origin of replication
- a modified ϕCTX integrase (*int*) gene
- a versatile multiple cloning site (MCS) flanked by T4 transcriptional termination sequences (Ω elements)
- the ϕCTX attachment site.

Several plasmids served as sources for the functional elements of the integration vectors mini-CTX1 and mini-CTX2. The plasmid backbone including the *tet* gene and *ori* was derived from pUCP26 (West et al.,

1994) (the N-terminal sequences of the *tet* gene located between the *Cla*I and *Nru*I sites were derived from pPS1010 to eliminate the *Eco*RV site) and the *oriT* (origin of transfer) was obtained from pGEM-T*oriT* (Schweizer et al., 1995). The *FRT* sequences, MCS, and Ω transcriptional terminators were derived by combining sequences from pPS854 (Hoang et al., 1998), pWSK29 (Wang and Kushner, 1991), and pACV (Schweizer and Hoang, 1995), respectively. Finally, p1000 and pIBH served as sources for the ϕCTX *attP* site and *int*, respectively (Wang et al., 1995). The *Eco*RI, *Eco*RV, *Kpn*I, *Pst*I, *Sma*I, and *Xho*I sites found in the *int* gene were eliminated by site-specific mutagenesis using the Altered Sites mutagenesis kit and CTX *int* mutagenic primers introducing single base changes while not altering the integrase amino acid primary sequence. The segment containing the $lacI^q$, P_{trc}, and *ori* of mini-CTX2 was excised from pSE380 (Brosius, 1989) on a 2652-bp *Nco*I–*Rca*I fragment.

The mini-CTX-T7 vector was derived in two steps. First, a 4419-bp fragment containing *lacI,* P_{lacUV5}, and the T7 *pol* was PCR amplified from pEB1 template DNA and subcloned into the *Bam*HI site of pUC19 after digestion with *Bam*HI. Second, the *lacI*-P_{lacUV5}-T7 *pol* gene cassette was subcloned into mini-CTX2 as an *Eco*RI–*Hin*dIII fragment to form mini-CTX-T7. The *asd*-based expression vector pPS1032 was constructed by ligation of a 4214-bp *Ase*I–*Hin*dIII fragment derived from pUCP20T*asd* to a 449-bp *Ase*I *Hin*dIII fragment from pET-15b. This generated an expression vector that contained *asd* as sole selectable marker, an *ori* for maintenance in *E. coli,* an *ori*1600 and Rep for replication in *P. aeruginosa,* an *oriT* for interspecies conjugal transfer, and the T7 promoter and His6-encoding domain with unique cloning sites.

The source of the *lasB–lacZ* transcriptional fusion used in this study was pLPBL. A *rhlA–lacZ* transcriptional fusion was derived from pLPBL in several steps. First, a *Bgl*II site was introduced upstream of the *lasB* regulatory region on pLPBL by tailing an *Eco*RI site with a nonphosphorylated *Bgl*II linker (5' CAGATCTGAATT 3') (Schweizer, 1992); this procedure resulted in pPS959, which contains a unique *Bgl*II site flanked by two *Eco*RI sites. Second, the region containing the *rhlA* regulatory region (Pearson et al., 1997) was excised from pUO58-24 on a

623-bp *Bgl*II–*Bam*HI fragment and cloned into pLITMUS29 to form pPS963. Finally, a *rhlA–lacZ* transcriptional fusion was derived by replacing a 434-bp *Bgl*II–*Bam*HI fragment on pPS959 containing the *lasB* regulatory region with a 624-bp *Bgl*II–*Hin*dIII yielding pPS964. For transfer into the *P. aeruginosa* genome, the *lasB–lacZ* and *rhlA–lacZ* fusions were then subcloned between between the *Eco*RI and the *Eco*RV sites of mini-CTX1 on; 3.8-kb *Eco*RI–*Dra*I fragments (the *Dra*I site is located 59 bp downstream of *lacZ* early in *lacY*), yielding pPS1041 and pPS1052, respectively.

DEVELOPMENT OF TOOLS FOR GENETIC MANIPULATION OF *Pseudomonas aeruginosa*

Suh et al. (2004) has developed several tools in order to facilitate study of the opportunistic bacterial pathogen, *Pseudomonas aeruginosa*. These tools included a series of cassettes carrying:

- The minimal sequence for the origin of transfer (*oriT*) of RP4 plasmid for introducing plasmid into *P. aeruginosa via* conjugation.
- A minimal sequence for *P. aeruginosa* replicon (stabilizing fragment) for maintenance of plasmids in *P. aeruginosa*.
- The transcriptionally non-polar tetracycline resistance gene (Tc^R) for insertional mutagenesis.
- Two conjugative and suicide lacZ reporter fusion plasmids for studying gene expression at the transcriptional and translational levels.
- A gentamicin resistant promoter-probing mini-Tn5*lacZ*.
- A tightly regulated T7 promoter/repressor system to control gene expression in *P. aeruginosa*.

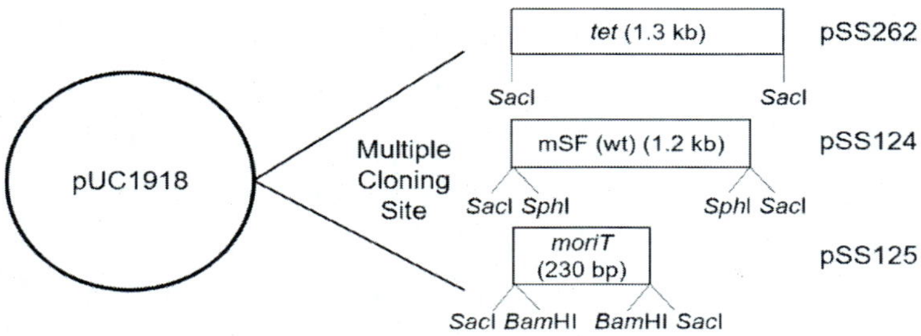

Figure 1.2. Genetic cassettes. A genetic cassette carrying the non-polar Tc^R gene can be isolated with *Sac*I, *Kpn*I, *Sma*I, *Xba*I, or *Pst*I from plasmid pSS262. The cassettes carrying either the m*oriT* or mSF can be isolated with any of the restriction enzymes present in the MCS of pUC1918. The m*oriT* cassette carries an additional *Bam*HI site internal to the *Sac*I site, while the mSF carries an additional *Sph*I site internal to the *Sac*I site.

The following genetic selection and screen procedures were used to facilitate clone isolation after ligation of a DNA fragment to the pUC1918-derived vector (Schweizer, 1993) (Figure 1.2). To isolate the plasmid carrying the m*oriT*, the ligation mixture was first electroporated into *E. coli* strain DH10B (Sm^R) and selected for Ap^R, the ampicillin resistance carried by the vector plasmid. The Ap^R colonies were then pooled and conjugated into *E. coli* strain DH5aF' (nalidixic acid resistance, Nal^R) *via* triparental mating with HB101/pRK2013 as the helper strain. The transconjugants were then selected for Ap^R (plasmid vector) and Nal^R (DH5aF') to counter-select against donors without the m*oriT* inserts (DH10B) and the helper strain (HB101). Plasmids were isolated from Ap^R Nal^R colonies and the presence of the m*oriT* was verified by restriction digest and PCR amplification. To isolate plasmids carrying the mSF, the ligation mixture was first electroporated into *E. coli* DH10B. The resulting Ap^R colonies were pooled and plasmids were isolated from the pool and electroporated into *P. aeruginosa* to select for those that can replicate autonomously in *P. aeruginosa*. Plasmids were then isolated from *P. aeruginosa* to verify presence of the mSF *via* restriction digest and PCR amplification.

In the construction of a plasmid that carried a tightly regulated promoter system, pSS213, the lac promoter, P_{lac}, in pSU38 (Martinez et al.,

1988) was replaced with a tightly regulated promoter, $P_{T7(A1/04/03)}$, from pUHE21-2 (Lanzer and Bujard, 1998). Briefly, the unique *Nhe*I site, located upstream of the P_{lac} in pSU38 was converted to a unique *Xho*I site by digesting with *Nhe*I, blunting the ends *via* fill-in reaction with Klenow, and ligating a *Xho*I linker. Subsequent digest with *Eco*RI in the MCS of pSU38 removed P_{lac} as a *Xho*I–*Eco*RI fragment. $P_{T7(A1/04/03)}$ from pUHE21-2 was then cloned as a *Xho*I–*Eco*RI fragment to replace P_{lac}, while preserving the MCS of pSU38. Second, the *lacI*^{q} gene, originally isolated as an *Eco*RI fragment from pMF54 (Franklin et al., 1994), was converted to a *Xho*I fragment, by blunting the ends with mung bean nuclease, and ligating the *Xho*I linker to the blunted DNA fragment. Finally, the Km^R carried on pSU38 was removed with *Nco*I and *Bgl*II, and the ends were converted to blunt ends with Klenow fragment *via* a fill-in reaction. Then *aacCI* (gentamicin resistance, Gm^R), acquired as a *Sma*I fragment from pUCGM1 (Schweizer, 1993), was cloned into the blunted former *Nco*I/*Bgl*II site to confer Gm^R to facilitate selection of the plasmid in *P. aeruginosa*.

CONSTRUCTION OF PCON2(3), PCON2(3)-KAN AND PCON2(3)-ZEO

Most cloning vectors available for *Pseudomonas* hosts are shuttle plasmids that replicate in both *E. coli* and *Pseudomonas* due to the ease of manipulation of *E. coli* and the many reagents available for this bacterium (Choi et al., 2008). These vectors contain two different origins of replication (Schumann, 2008), in which *rep* (pMB1), the one active in *E. coli* and stabilizing fragment (SF) for *Pseudomonas* were utilised in this study. Besides a replicon for stable maintenance in respective hosts, other basic requirements include selection markers (kanamycin and zeocin) for identification of cells containing and maintaining the plasmid and restriction sites for cloning of heterogeneous DNA fragments. Versatility of vectors was introduced by integrating non-essential features such as genetic elements facilitating screening of recombinants, elements for interspecies transfer of recombinant plasmids and most importantly, promoters whose expression levels can be regulated and overexpressed.

Construction of *Escherichia-Pseudomonas* shuttle vectors involved the insertion of fragments from pUCP19, pSS213, pUCP19/HindIII1500PstI and pSS213/PstI1500HindIII. PstI1500HindIII/T7$_{(A1/O4/O3)}$, *kan/pro* and *EM7/Sh ble* are several genetic elements needed to be incorporated into pUCP19/HindIII1500PstI in order to form pCon2(3), pCon2(3)-Kan and pCon2(3)-Zeo, respectively. From Figure 1.3(A), PstI1500HindIII/ T7$_{(A1/O4/O3)}$ encoding the ORF of organic solvent tolerant elastase from *P. aeruginosa* strain K, MCS of pUC18 and modified T7 promoter, *T7*$_{(A1/O4/O3)}$, was successfully amplified with a size of approximately 1.6 kb. *kan/pro* and *EM7/Sh ble* which encoded their respective kanamycin and zeocin resistant genes, on the other hand, were generated with the size of about 0.9 [Figure 1.3(B)] and 0.45 kb [Figure 1.3(C)], respectively.

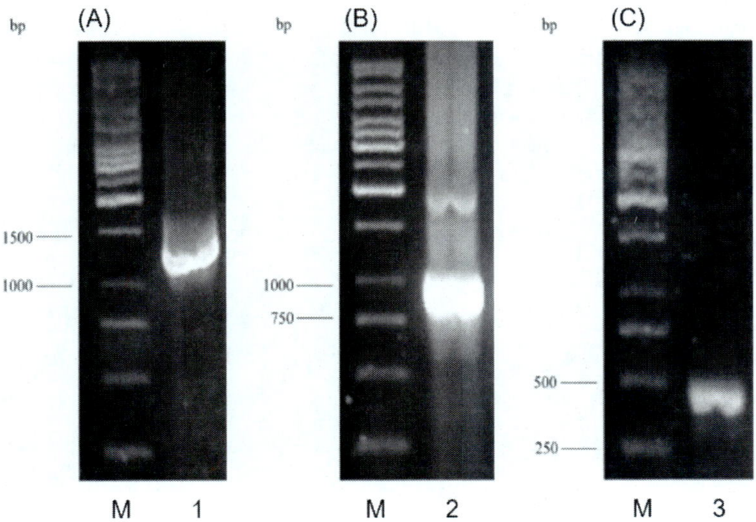

Figure 1.3. Amplification of genetic elements involved in 3 constructs. (A) DNA fragments of PstI1500HindIII/T7$_{(A1/O4/O3)}$ involved in construction of pCon2(3). (B) Kanamycin resistant gene, *kan/pro*, for construction of pCon2(3)-Kan. (C) PCR products of *EM7/Sh ble* in pCon2(3)-Zeo. Lane M: GeneRuler™ 1 kb DNA Ladder (Fermentas, Canada); Lane 1: PstI1500HindIII/T7$_{(A1/O4/O3)}$; Lane 2: *kan/pro*; Lane 3: *EM7/Sh ble*.

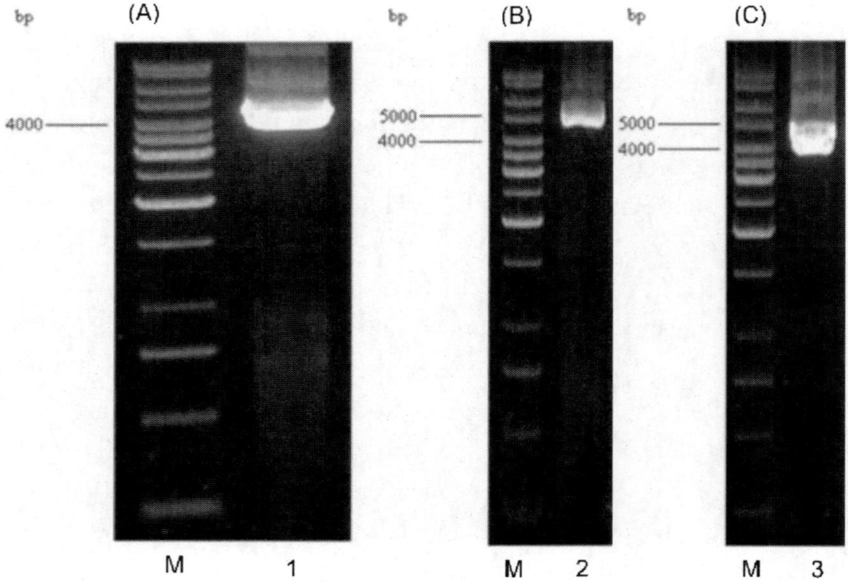

Figure 1.4. Extraction of constructed vectors from *E. coli* TOP10. (A) Insertion of PstI1500HindIII/T7$_{(A1/O4/O3)}$ from pSS213/PstI1500HindIII to pUCP19/HindIII1500PstI give rise to pCon2(3). (B) Derivation of pCon2(3), pCon2(3)-Kan, with additional kanamycin resistant gene as selection marker. (C) Another derivation of pCon2(3), pCon2(3)-Zeo, with additional zeocin resistant gene as selection marker. Lane M: GeneRuler™ 1 kb DNA Ladder (Fermentas, Canada); Lane 1: pCon2(3); Lane 2: pCon2(3)-Kan; Lane 3: pCon2(3)-Zeo.

Several positive clones of *E. coli* TOP10 harboring constructed vectors were selected randomly for plasmid analysis. Initial investigation demonstrated that plasmid from Lane 1 showed potential ligation of PstI1500HindIII/T7$_{(A1/O4/O3)}$ to pUCP19/HindIII1500PstI as an intense band with a size of 4.5 kb was observed in Figure 1.4(A). Plasmids from Figure 1.4(B) and Figure 1.4(C) exhibited similar observation as Lanes 2-5 and Lanes 6-9 which represented pCon2(3)-Kan and pCon2(3)-Zeo, respectively, clearly showed concentrated bands.

The construction of vectors from Figure 1.4 was further confirmed by *Nde*I- and *Psc*I-mediated digestion. In pCon2(3), restriction endonuclease *Nde*I will digest the constructed plasmid into two fragments due to the presence of *Nde*I linker in PstI1500HindIII/T7$_{(A1/O4/O3)}$ prior to ligation to pUCP19/HindIII1500PstI. Figure 1.5(A) clearly showed that PstI1500

HindIII/T7$_{(A1/O4/O3)}$ ligated successfully to pUCP19/HindIII1500PstI as both of the vectors exhibited their actual sizes of 6.0 and 1.6 kb, respectively, after digestion. Therefore, this construct was named pCon2(3) with 7.6 kb. Positive observation was also seen for the construction of pCon2(3)-Kan and pCon2(3)-Zeo. Both of these constructs completely released its antibiotic selection markers after digestion with *Psc*I due to the insertion of *kan/pro* and *EM7/Sh ble* at the *Psc*I site of pCon2(3). From Figure 1.5(B) and Figure 1.5(C), two bands with the sizes of 7.6 kb for pCon2(3) and 0.9 and 0.45 kb which represented *kan/pro* and *EM7/Sh ble*, respectively, were observed after digestion with *Psc*I. As the result, both pCon2(3)-Kan and pCon2(3)-Zeo were concluded to encode approximately 8400 and 8050 nucleotides, respectively.

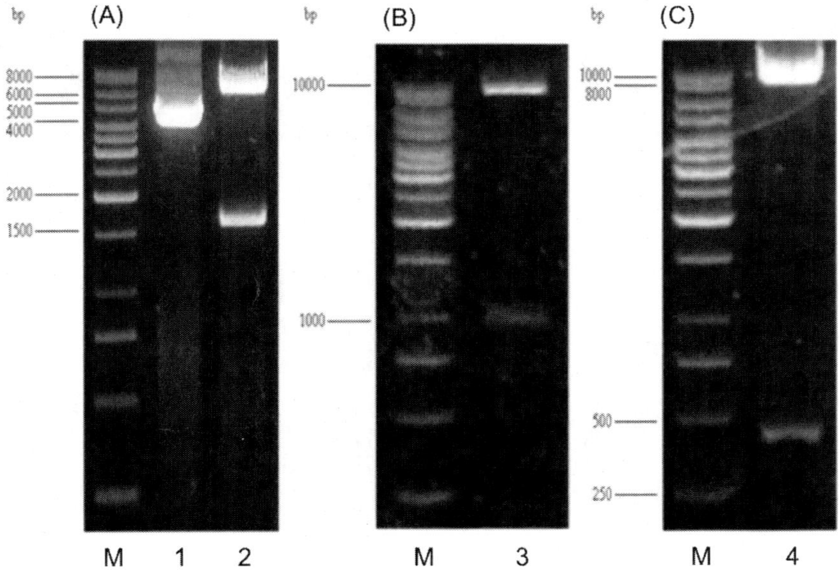

Figure 1.5. Confirmation of constructed vectors by restriction endonuclease digestion. (A) pCon2(3) released its *T7*$_{(A1/O4/O3)}$ controlled-expression cassette, PstI1500HindIII/T7$_{(A1/O4/O3)}$, after digestion with *Nde*I. (B) Digestion of pCon2(3)-Kan by *Psc*I resulted in two fragments, pCon2(3) and kan/pro. (C) The release of Sh/ble from pCon2(3)-Zeo was facilitated by *Psc*I. Lane M: GeneRuler™ 1 kb DNA Ladder (Fermentas, Canada); Lane 1: pCon2(3); Lane 2: pCon2(3) after digestion with *Nde*I; Lane 3: pCon2(3) after digestion with *Psc*I; Lane 4: pCon2(3)-Zeo after digestion with *Psc*I.

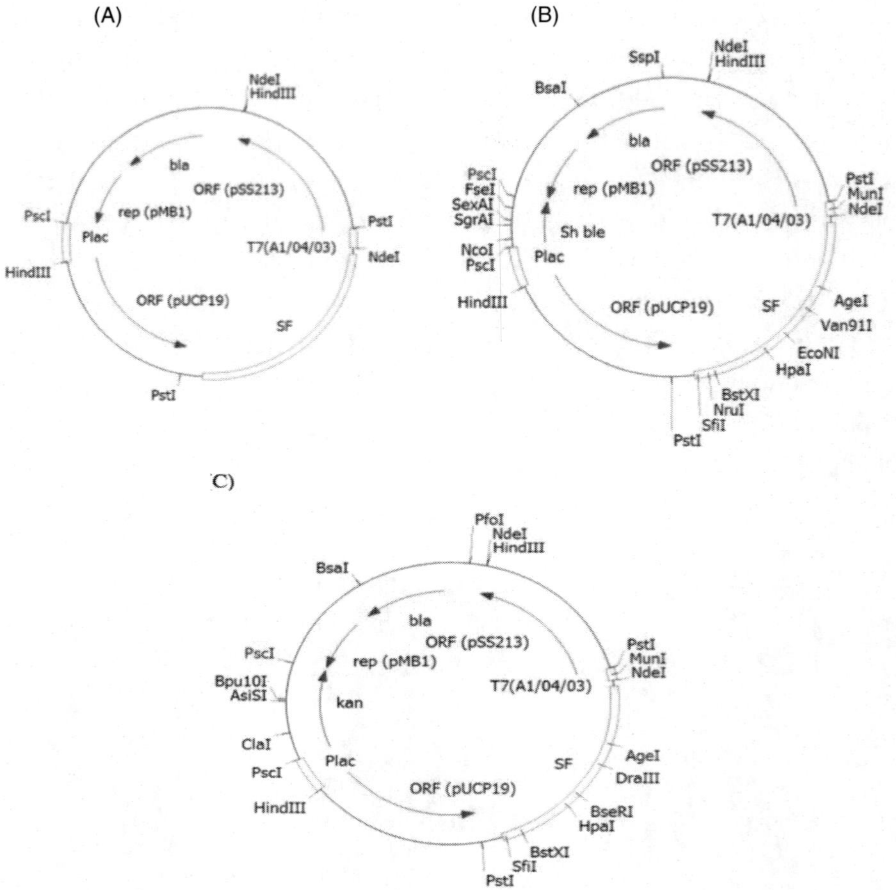

Figure 1.6. Physical maps of constructed shuttle vectors carrying P_{lac} and $P_{T7(A1/O4/O3)}$ expression cassettes. (A) pCon2(3), (B) pCon2(3)-Zeo and (C) pCon2(3)-Kan. The maps were generated by Clone Manager 9 Basic Edition (Scientific and Educational Software, USA) and not drawn according to scale.

New and novel *Escherichia-Pseudomonas* shuttle vectors, pCon2(3) [Figure 1.6(A)], pCon2(3)-Kan [Figure 1.6(B)] and pCon2(3)-Zeo [Figure 1.6(C)], were generated from several sources, namely pUCP19/HindIII 1500PstI, PstI1500HindIII/T7$_{(A1/O4/O3)}$, *kan/pro* and *EM7/Sh ble*, in which all of the DNA fragments were originated from pUCP19 (Schweizer, 1991), pSS213 (Suh *et al.,* 2004), pSTBlue-1 (Novagen, Germany) and pPICZαA (Invitrogen, USA), respectively. Both of the pUCP19 and pSS213 carried the organic solvent tolerant elastase gene from *P.*

aeruginosa strain K. pSS213 with the size of 3.8 kb, is a versatile vector that carried gentamicin resistant gene, Gm^R, for selection, a p15A replicon for maintenance in *E. coli*, the multiple cloning sites (MCS) of pUC18 for cloning, the modified phage T7 early gene promoter, $T7_{(A1/O4/O3)}$ for control of gene expression, and the $lacI^q$ gene for repression of the $T7_{(A1/O4/O3)}$ promoter.

In 1988, Lanzer and Bujard had discovered that kinetic parameters of RNA polymerase-promoter interaction as well as the position of an operator within the promoter sequence drastically affect the occupancy of the operator by its repressor, which ultimately defines the efficiency of repression. The construction of *O3* (a sequence that contained 29 bp wild type operator in a position homologous to P_{lac}) and *O4* (sequence that carried a 17 bp core region of the wild type *lac* operator as spacer between -10 and -33 hexamers of the promoter) into P_{A1}, a promoter derived from coliphage T7, had evidenced in high rate of promoter clearance which lower the occupancy of the promoter by RNA polymerase, thereby increases the repression factor. Therefore, the $T7_{(A1/O4/O3)}$ promoter is reported to be repressed tightly because it contains two *lac* operator sites for binding of Lac repressor expressed by $lacI^q$ (Lanzer and Bujard, 1988), but addition of IPTG relieves repression and the gene of interest is expressed.

Expression vectors are designed to achieve regulated expression of cloned genes from inducible promoters. The corresponding regulatory genes are either contained on the same plasmid or are provided *in trans*, either encoded by a compatible plasmid or a chromosomally integrated element (Choi et al., 2008). Among many combinations of inducible promoter/regulatory elements that have been used, the *lac* operon-based promoter including P_{lac} and the T7 promoter (P_{T7}) are probably the two that are most easily to adapt for use especially in pseudomonads and *E. coli* because their biology is well-understood and the molecular reagents required for establishment of these expression systems are readily available in most laboratories. The constructed vectors, as documented earlier, encoded double expression cassettes from pUCP19/HindIII1500PstI and pSS213/PstI1500HindIII, with each controlled by P_{lac} and $P_{T7(A1/O4/O3)}$, respectively. Thus, the transcription and translation processes of the desired gene will be aimed to be twice as active as the native vectors, thereby predicting the expression fold to be at least two times more than the unmodified plasmids.

These vectors visualized the functionality of loosely regulated *lac* promoter and tightly regulated T7 promoter/repressor system in controlling the expression level. In earlier experiments, Suh *et al*. (2004) has placed a quorum sensing gene regulator *rhlR* under control of the $T7_{(A1/O4/O3)}$ promoter system in pSS213. In the absence of induction, expression of *rhlR* under control of the $T7_{(A1/O4/O3)}$ promoter produced a phenotype that resembled that of a *rhlR* null mutant strain, as indicated by the lack of elastase and pyocyanin production. When 1 mM IPTG was added, *rhlR* under the $T7_{(A1/O4/O3)}$ promoter was expressed and the phenotype resembled that of the wild-type strain of *P. aeruginosa* PA01.

Nucleotide Sequence Analysis of pCon2(3), pCon2(3)-Kan and pCon2(3)-Zeo

Nucleotide sequence of pCon2(3) can be retrieved from GenBank® accession no. JF276396. The structural gene of elastase strain K controlled by $P_{T7(A1/O4/O3)}$ (PstI1500HindIII/ $P_{T7(A1/O4/O3)}$), which inserted at *Nde*I of position 300 and 1941, is located at positions 314-1810, with 1810 being the translational start sequence of the protein. The promoter -10 (AGTATC) and -33 (AGTCAA) boxes were found at respective 81 and 104 bp downstream of position 1810. The operon of $P_{T7(A1/O4/O3)}$ was claimed to consist of two operators (Suh et al., 2004). In respect to this, operator *O3* (CTGTGTGAAATTGTTATCCGCTCACAATT) is observed at positions 1856-1884, 7 bp upstream of -10 box. Meanwhile, operator *O4* (ATTGTTATCCGCTCACA) is situated in between the -10 and -33 boxes. The stabilizing fragment (SF) of 1881 bp, which confers stability and plasmid maintenance in *Pseudomonas*, is seen to span across positions 1998-3879.

Table 1.1. Genetic elements in constructed vectors

Source	pCon2(3)	pCon2(3)-Kan	pCon2(3)-Zeo	pCon4	pCon5
pUCP19	Expression cassette controlled by P_{lac}	Expression cassette controlled by P_{lac}	Expression cassette controlled by P_{lac}	-	-
	Stabilizing fragment	Stabilizing fragment	Stabilizing fragment	Stabilizing fragment	Stabilizing fragment
	Carbenicillin resistance	Carbenicillin resistance	Carbenicillin resistance	Carbenicillin resistance	Carbenicillin resistance
	ColE1 origin of replication	ColE1 origin of replication	ColE1 origin of replication	ColE1 origin of replication	ColE1 origin of replication
pSS213	Expression cassette controlled by $P_{T7(A1/04/03)}$	Expression cassette controlled by $P_{T7(A1/04/03)}$	Expression cassette controlled by $P_{T7(A1/04/03)}$	Expression cassette controlled by $P_{T7(A1/04/03)}$	Expression cassette controlled by $P_{T7(A1/04/03)}$
	-	-	-	-	*lacI^q^*
pSTBlue-1	-	Kanamycin resistance	-	-	-
pPICZαA	-	-	Zeocin resistance	-	-

'-' Absence of genetic tools in constructed vector.

The second expression cassette was constructed by insertion elastase strain K structural gene (C- and N-terminal at 4072 and 5568, respectively) at *Pst*I and *Hin*dIII sites of P_{lac} multiple cloning sites. The expression of this fragment is initiated by -10 (ATTCC) and -30 (TGTAAA) boxes, both located at positions 5628-5632 and 5661-5666, respectively. The fragment which responsible for plasmid replication, *rep* (pMB1), is positioned at 5975-6593. Flanking the region of 6749-7618 is the *bla* gene that coded for resistance to β-lactam group of antibiotics, is found near the 3'-end of pCon2(3) sequence. While preserving the nucleotide sequence positions of fragments, starting from upstream of P_{lac} promoters of pCon2(3), the positions of *rep* (pMB1) and *bla* had changed to respective 6424-7042 and 7198-8067 in pCon2(3)-Zeo since the insertion of 442 bp zeocin resistant gene, *EM7/Sh ble*, at *Psc*I site (5929, 6378) of pCon2(3) (GenBank® accession no. JF276393). Similarly, *kan/pro*, a gene for kanamycin resistant, was incorporated into *Psc*I site (5929, 6896) of pCon2(3) to give rise to pCon2(3)-Kan and thus resulted in the shift of *rep* (pMB1) and *bla* to 6942-7560 and 7716-8585, respectively (GenBank® accession no. JF276397). Summary of putative genetic elements present in pCon2(3), pCon2(3)-Kan and pCon2(3)-Zeo is listed in Table 1.1.

Construction of pCon4 and pCon5

Two genetic tools, pCon4 and pCon5, were developed in order to facilitate the study of protein expression in different promoter/repressor system by opportunistic bacterial pathogen *P. aeruginosa* and *E. coli* (Figure 1.7). An intermediate vector pLac$^-$ with a size of approximately 4.2 kb was constructed prior to development of pCon4 and pCon5 (data not shown). The absence of MCS of pUCP19 and lac promoter in pLac$^-$ will greatly assist in promoter/repressor substitution by modified T7 promoter, $T7_{(A1/O4/O3)}$, MCS of pUC18 and lac repressor, *lacI^q*, in pCon4 and pCon5.

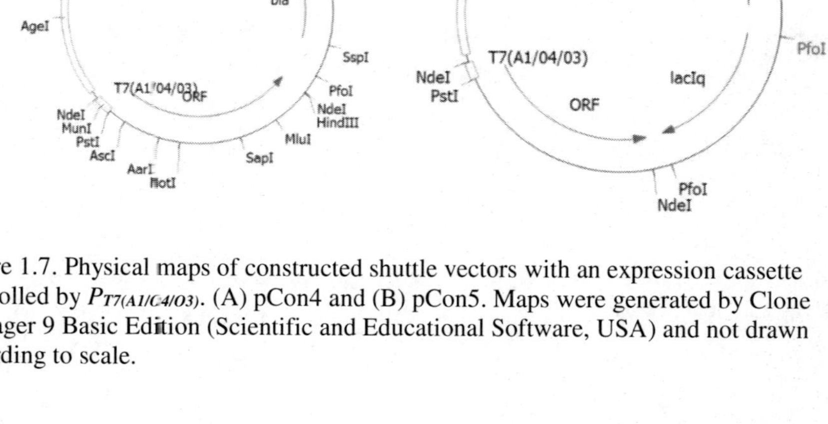

Figure 1.7. Physical maps of constructed shuttle vectors with an expression cassette controlled by $P_{T7(A1/O4/O3)}$. (A) pCon4 and (B) pCon5. Maps were generated by Clone Manager 9 Basic Edition (Scientific and Educational Software, USA) and not drawn according to scale.

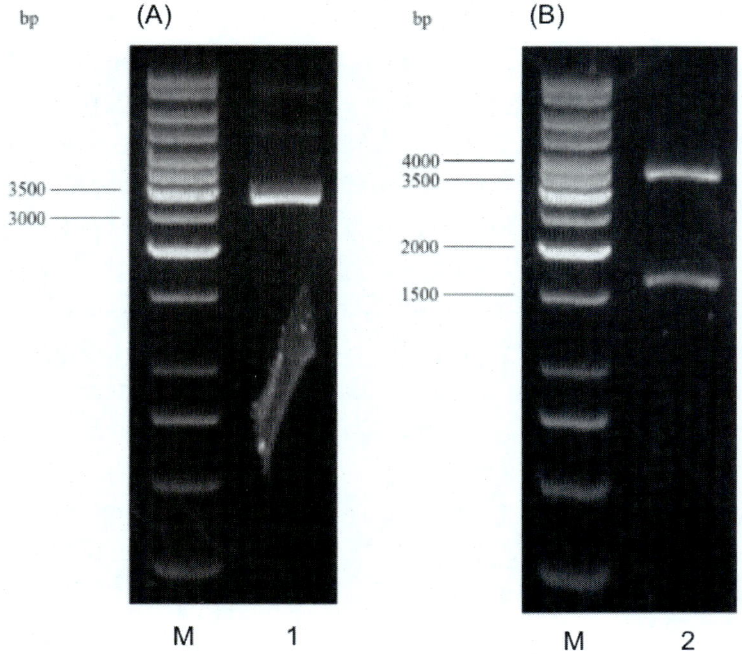

Figure 1.8. Gel electrophoresis from recombinant *E. coli* TOP10/pCon4. (A) Extraction of pCon4 from *E. coli* TOP10/pCon4. (B) Single digestion of pCon4 by *Nde*I. Lane M: GeneRuler™ 1 kb DNA Ladder (Fermentas, Canada); Lane 1: pCon4; Lane 2: *Nde*I digested-pCon4.

Figure 1.8(A) visualized the presence of constructed plasmid, pCon4, with the size of 3.0 kb from recombinant *E. coli* TOP10. Restriction endonuclease digestion by *Nde*I, a site of insertion of PstI1500HindIII/ $T7_{(A1/O4/O3)}$ to pLac$^-$, illustrated the release of PstI1500HindIII/$T7_{(A1/O4/O3)}$ (1.6 kb) which harboured the $T7_{(A1/O4/O3)}$, ORF of organic solvent tolerant elastase from *P. aeruginosa* strain K and MCS of pUC18 from pCon4 [Figure 1.8(B)]. Therefore, pCon4 which aimed to visualize the functionality of tightly regulated T7 promoter system in controlling the expression level, was deduced to encode 5.8 kb of nucleotides.

Detection of construct pCon5 involved the selection of several recombinant clones from *E. coli* TOP10/pCon5. Majority of the extracted plasmids from Lanes 1-5, 7-8 and 10 demonstrated a size similar to that of pCon4, which was around 3.0 kb [Figure 1.9(A)]. In contrast, Lanes 6 and 9 in Figure 1.9(A) exhibited a larger size of 4.0 kb. Further confirmation by *Pfo*I digestion also showed that only plasmids from Lanes 16 and 19 (which originated from respective Lanes 6 and 9) contained the lac repressor gene, *lacI^q^*, with the size of 1.2 kb [Figure 1.9(B)]. This result was in agreement with Section 3.2.22.5 as *Pfo*I site was the point insertion of $lacI^q$ into pCon4 during the construction of pCon5. Vectors from Lanes 11-15, 17-18 and 20 in Figure 1.9(B), however, showed false-positive results due to recirculization of pCon4 resulted by incomplete dephosphorylation of pCon4 prior to ligation to $lacI^q$. Construction of pCon5 will display the repression level of $lacI^q$ on $T7_{(A1/O4/O3)}$ promoter. This construct, in fact, will reduce the level of basal transcription or leakiness that typically occurred in *lac*-mediated promoter (absence of induction) in *Pseudomonas* sp. and *E. coli* (Schweizer, 2001).

Additionally, high level of repression also can be achieved by the addition of a copy of the *lacI* and/or T7 lysozyme gene (gene *3.5)* directly onto vectors, especially important for the cloning and expression of potentially toxic gene products in *P. aeruginosa* (Brunschwig and Darzins, 1992). However, a sufficient amount of Lac repressor has to be synthesized in order to obtain strong repression in the absence of inducer. Since the wild type promoter of *lacI* gene is rather weak (10-20 molecules of Lac repressor per wild type cell), a promoter-up mutation is used which designated as $lacI^q$ where 'q' stands for quantity. Here, about 100 repressor molecules are present per cell if the $lacI^q$ allele is present in the chromosome (Schumann, 2008). Table 1.1 summarizes the presence of putative genetic elements in each constructed expression vectors of pCon4 and pCon5.

Nucleotide Sequence Analysis of pCon4 and pCon5

From the sequence obtained from GenBank® accession no. JF276394, origin of replication for *E. coli*, *rep* (pMB1), is located at positions 47-665 in pCon4 nucleotide sequence. The *bla* gene (821-1690), on the other hand, is observed at position 155 bp downstream of *rep* (pMB1). At the *Nde*I of positions 2064 and 3705, the structural gene of elastase strain K (as a passenger enzyme) with translational start site at 3574 was inserted together with $T7_{(A1/O4/O3)}$ operon that encodes for -10 box, -33 box, operator *O3* and operator *O4* spanning at nucleotides 3656-3660, 3679-3684, 3619-3648 and 3662-3678, respectively. The *Pseudomonas* stabilizing fragment (SF) is seen to flank the region of 3762-5643 in pCon4. Sequence of pCon5 is differed in respect to pCon4 by the presence of 1229 bp of *lac* repressor gene, *lacI*^{q}, at *Pfo*I site (1927, 3176) (GenBank® accession no. JF276395). Meanwhile, the nucleotide positions of *rep* (pMB1) and *bla* in pCon5 are similar with the one in pCon4. Due to the addition of *lacI*^{q} in pCon5, the $T7_{(A1/O4/O3)}$ operon including -10 box, -33 box, operator *O3*, operator *O4* as well as SF region had shifted, each at positions 4905-4910, 4928-4933, 4869-4897, 4911-4927 and 5011-6892.

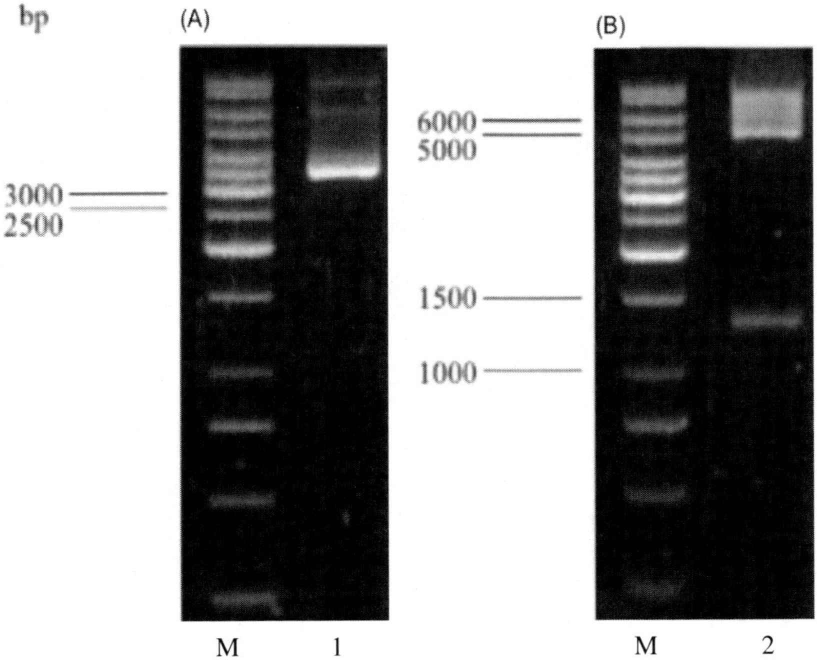

Figure 1.9. Gel electrophoresis from recombinant *E. coli* TOP10/pCon5. (A) Extraction of pCon5 from *E. coli* TOP10/pCon5. (B) Single digestion of pCon5 by *Pfo*I. Lane M: GeneRuler™ 1 kb DNA Ladder (Fermentas, Canada); Lane 1: pCon5; Lane 2: *Pfo*I digested-pCon5.

Protein Expression from pCon2(3), pCon2(3)-Kan, pCon2(3)-Zeo, pCon4 and pCon5

All constructed shuttle vectors were transformed into various strains of *E. coli* and *P. aeruginosa* to determine the best elastinolytic expression level exhibited by the vectors in multiple hosts. In general, *E. coli* TOP10 was able to replicate and express the heterogeneous protein harbored by all constructed vectors (Figure 1.10). Construct pUCP19/HindIII1500PstI, which carried single expression cassette controlled by P_{lac}, showed the highest protease activity of 4.24 A_{495}/h/mL with an increase of 16.64 in TOP10 compared to control (pUCP19). Although exhibited high activity, constructs which harbored two expression cassettes such as pCon2(3), pCon2(3)-Kan and pCon2(3)-Zeo were unable to exhibit greater protease activity due to the fact that *E. coli* TOP10 is incapable of utilizing $T7_{(A1/O4/O3)}$ promoter for protein expression. This was also evidenced in TOP10/pCon4 and TOP10/pCon5 that controlled only by $T7_{(A1/O4/O3)}$ promoter, in which slight increase in activities was observed. Among the constructs with double expression cassettes in TOP10, pCon2(3)-Kan had demonstrated the best expression of 3.56 A_{495}/h/mL, an increase of 13.97-fold against its control.

KRX is an *E. coli* K12 derivative that has salient features associated with cloning and screening strains plus engineered attributes to optimize controlled protein expression. KRX incorporates a chromosomal copy of the T7 RNA polymerase gene driven by the rhamnose promoter (*rha*PBAD) to provide dramatic control of recombinant protein expression in P_{T7} promoter. One important reason for the success of this system is that the recombinant genes are fused to the very strong T7-promoter, which is not recognized by *E. coli* RNA polymerase. Therefore, virtually no expression is observed until the gene coding for T7 RNA polymerase is expressed. The T7 RNA polymerase consists of only single polypeptide chain of approximately 100 kDa, in which possesses a high processivity that initiates RNA chains very efficiently and elongates them about five times faster than *E. coli* enzyme and it is rifampicin resistant (Schumann, 2008).

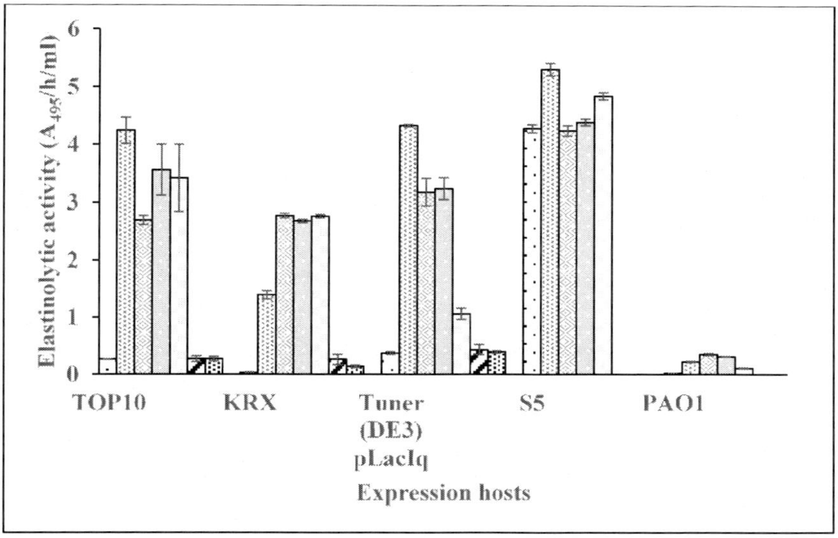

Figure 1.10. Expression analysis of constructed vectors in various strains of *E. coli* and *P. aeruginosa*. The tested vectors are pUCP19 ( ) as negative control, pUCP19/HindIII1500PstI ( ) as positive control, pCon2(3) ( ), pCon2(3)-Kan ( ), pCon2(3)-Zeo ( ), pCon4 ( ) and pCon5 ( ). Elastinolytic activities were detected from intracellular fraction of *E. coli* strains while *Pseudomonas* strains exhibited extracellular activities. Values are represented by mean value ± standard deviations (n=3) (error bars), which obtained after 6 h of induction with 1.0 mM IPTG.

Overexpression of elastase strain K was viewed on *E. coli* KRX/pCon2(3) (Figure 1.10). The latter displayed greatest elastinolytic activity and fold of 2.77 A_{495}/h/mL and 81.50, respectively, compared to control. Similar overexpression pattern were obtained for KRX/pCon2(3)-Kan and KRX/pCon2(3)-Zeo with their respective activities of 2.68 and 2.76 A_{495}/h/mL. The former and latter constructs had yielded respective fold of 78.85 and 81.29 in respect to its controls. With the presence of extra *lac* repressor gene, KRX/pCon5 had recorded a lower activity of 0.14 A_{495}/h/mL compared to KRX/pCon4 (0.26 A_{495}/h/mL), which possessed only low amount of repressor from bacterial genome. Elastinolytic activity in KRX/pUCP19/HindIII1500PstI, in contrast, had recorded approximately 50% of its counterparts with double expression cassettes under the same host. Meanwhile, pUCP19/HindIII1500PstI in *E. coli* Tuner™ (DE3) pLacI displayed the greatest expression of elastase activity of 4.33

A_{495}/h/mL compared to other constructs transformed under the same host with 11.76 increase in fold. Increment of activities was also viewed on pCon2(3), pCon2(3)-Kan and pCon2(3)-Zeo by 8.64-, 8.81- and 2.88-fold, respectively, compared to their controls. From Figure 1.10, a similar observation was detected for construct pCon4 and pCon5, each with elastinolytic activity of 0.43 and 0.39 A_{495}/h/mL, which carried single expression cassette controlled by *T7*$_{(A1/O4/O3)}$ promoter in *E. coli*, whereby both constructs did show significant expression level but lower than that of constructs with two expression cassettes. This might be due to the presence of two *lac* operator sites for binding of Lac repressor that conferred to tight repression of the promoter in addition to the presence of *lac* repressor gene carried by plasmid Tuner™ (DE3) pLacI and pCon5. However, high expression level was observed when the *lac* and *T7*$_{(A1/O4/O3)}$ promoters appeared simultaneously in *E. coli*.

Unlike *E. coli*, only several constructs namely pUCP19/HindIII1500 PstI, pCon2(3), pCon2(3)-Kan and pCon2(3)-Zeo were able to replicate in *P. aeruginosa* S5 and PAO1. Similar to the expression of recombinant elastase strain K in *P. aeruginosa* as reported earlier, elastinolytic activities exhibited by these constructs were also detected at the extracellular culture media of S5 and PAO1. With exception given to pUCP19/HindIII1500PstI and pCon2(3)-Zeo of S5 which showed significant expression level of 5.31 and 4.85 A_{495}/h/mL, respectively, the elastinolytic activities exhibited by S5/pCon2(3) and S5/pCon2(3)-Kan, however, were comparable to that of their controls (Figure 1.10).

As a comparison, *P. aeruginosa* PAO1/pCon2(3) had overexpressed and recorded an elastinolytic activity of 0.35 A_{495}/h/mL with greater fold of 14.64 in relative to its control (Figure 1.10). Additionally, other PA01 constructs such as pUCP19/HindIII1500PstI, pCon2(3)-Kan and pCon2(3)-Zeo had also documented respective substantial expression level of 0.22, 0.31 and 0.11 A_{495}/h/mL. Suh *et al.* (2004) had tested the efficiency of *T7*$_{(A1/O4/O3)}$ by placing a promoterless and 3’ truncated autoinducer for elastase and pyocynin, *rhlR’*, into pSS213 for intergration into *P. aeruginosa* PAO1 genome by homologous recombination of *rhlR* sequences in addition to *moriT* to allow mobilization into *P. aeruginosa via* conjugation. This resulted in an elastase activity of 102.5 A_{495}/min/g from 86.7 A_{495}/min/g (control), an increment of 18.0% following induction at 1.0 mM of IPTG. Under the same host, the construct pCon2(3), therefore, was capable to overexpress greater elastinolytic fold with an increment of 1464.4% in respect to the one reported by Suh and colleagues.

Circular conformation of pCon2(3), pCon2(3)-Zeo and pCon2(3)-Kan was extracted from *P. aeruginosa* strain PAO1 and S5 with the size of approximately 10.0-6.0 kb (Figure 1.11). Comparable to pUCP19/ HindIII1500PstI (data not shown), the yield of the constructs replicated from PAO1 and S5 were qualitatively low compared the ones found in strains of *E. coli*. This result had indicated that the copy number of pUCP19-derivatives was still remained unchanged, although multiple insertions of DNA fragments were carried out on the native plasmid during vector construction. The pCon4 and pCon5, however, were unable to replicate in both PAO1 and S5 as desired bands were undetected following plasmid extraction (Figure 1.11).

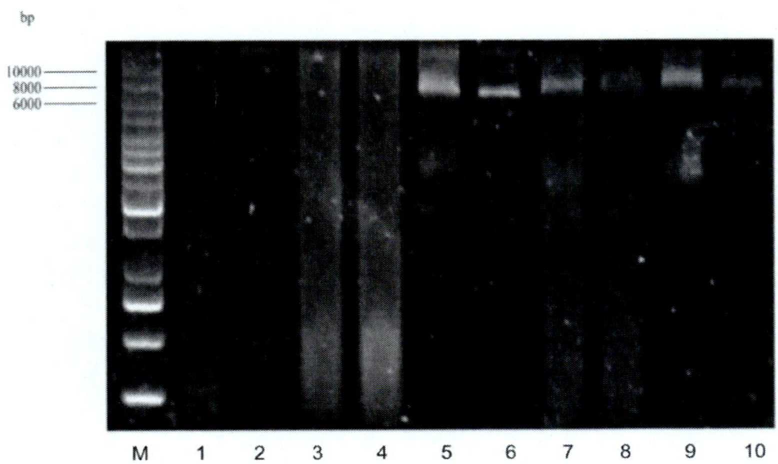

Figure 1.11. Detection of constructs from PA01 and S5. GeneRuler™ 1 kb DNA Ladder (Fermentas, Canada) is represented by lane M. Constructed vectors of pCon4, pCon5, pCon2(3), pCon2(3)-Zeo and pCon2(3)-Kan are loaded into lane 1-2, 3-4, 5-6, 7-8 and 9-10, respectively. Samples of odd numbered lanes are extracted from PAO1 while even numbered lanes are originated from S5.

References

Brosius, J. (1989). Superpolylinkers in Cloning and Expression Vectors. *DNA*, 8, pp. 759-777.

Brunschwig, E. and Darzins, A. (1992). A two-component T7 System for the Overexpression of Genes in *Pseudomonas aeruginosa*. *Gene*, 111, pp. 35-41.

Chen, Z., Chen, H., Ni, Z., Tian, R., Zhang, T., Jia, J. et al. (2015). Expression and Characterization of A Novel Nitrilase from Hyperthermophilic Bacterium *Thermotoga maritime* MSB8. *Journal Microbiology and Biotechnology,* 25, pp. 1660-1669.

Choi, K., Trunck, L. A., Kumar, A., Mima, T., Karkhoff-Schweizer, R. R. and Schweizer, H. P. (2008). Genetic Tools for *Pseudomonas*. In: Cornelis, P., ed., *Pseudomonas Genomics and Molecular Biology*, Norfolk: Caister Academic Press, pp: 65-86.

Cook, T. B., Rand, J. M., Nurani, W., Courtney, D. K., Liu, S. A. and Pfleger, B. F. (2018). Genetic Tools for Reliable Gene Expression and Recombining in *Pseudomonas putida. Journal of Industrial Microbiology & Biotechnology,* 45, pp. 517-527.

Dammeyer, T., Steinwand, M., Kruger, S. C., Dűbel, S., Hust, M. and Timmis, K. N. (2011). Efficient Production of Soluble Recombinant Single Chain Fv Fragment by a *Pseudomonas putida* strain KT2440 Cell Factory. *Microbial Cell Factories,* 10, pp. 11.

Franklin, M. J., Chitnis, C. E., Gaccesa, P., Sonesson, A., White, D. C. and Ohman, D. E. (1994). *Pseudomonas aeruginosa* AlgG is a Polymer Level Alginate C5-mannuronan Epimerase, *Journal of Bacteriology*, 176, pp. 1821-1830.

Hoang, T. T., Karkhoff-Schweizer, R. R., Kutchma, A. J. and Schweizer, H. P. (1998). A broad-host-range Flp-*FRT* Recombinant System for Site Specific Excision of Chromosomally Located DNA Sequences: Application for Isolation of Unmarked *Pseudomonas aeruginosa* Mutants. *Gene*, 212, pp. 77-86.

Hoang, T. T., Kutchma, A. J., Becher, A. and Schweizer, H. P. (2000). Integration-proficient plasmids for *Pseudomonas aeruginosa*: Site-specific Integration and Use for Engineering of Reporter and Expression Strains. *Plasmids*, 43, pp. 59-72.

Huang, W. and Wilks, A. (2017). A Rapid Seamless Method for Gene Knockout in *Pseudomonas aeruginosa. BioMed Central Microbiology,* 17, pp. 199.

Joshi, H., Dave, R. and Venugopalan, V. P. (2014). Protein as Chemical Cue: Non Nutritional Growth Enhancement by Exogenous Protein in *Pseudomonas putida* KT2440. *PLoS ONE* 9, pp. e103730.

Lanzer, M. and Bujard, H. (1998). Promoters Largely Determine the Efficiency of Repressor Action. *Proceedings of the National Academy of Sciences of the United States of America*, 85, pp. 8973-8977.

Liu, W., Li, M. and Yan, Y. (2017). Heterologous Expression and Characterization of a New Lipase from *Pseudomonas fluorescens* Pf0-1 and Used for Biodiesel Production. *Scientific Reports,* 7, pp. 15711.

Loeschcke, A. and Thies, S. (2015). *Pseudomonas putida*- A Versatile Host for the Production of Natural Products. *Applied Microbiology and Biotechnology,* 99, pp. 6197-6214.

Martinez, E., Bartolome, B. and de la Cruz, L. (1988). pACY184-derived Cloning Vectors Containing the Multi Cloning Site and *lacZα* Reporter Gene of pUC8/9 and pUC18/19 Plasmids. *Gene*, 68, pp. 159-162.

Minton, N. P., Atkinson, T. and Sherwood, R. F. (1983). Molecular Cloning of the *Pseudomonas* carboxypeptidase G_2 Gene and Its Expression in *Escherichia coli* and *Pseudomonas putida. Journal of Bacteriology*, 156, pp. 1222-1227.

Pasloske, B. L., Drummond, D. S., Frost, L. S. and Paranchych, W. (1989). The Activity of the *Pseudomonas aeruginosa* Pilin Promoter is Enhanced by An Upstream Regulatory Site. *Gene*, 81, pp. 25-34.

Pearson, J. P., Pesci, E. C. and Iglewski, B. H. (1997). Roles of *Pseudomonas aeruginosa las* and *rhl* Quorum Sensing Systems in Control of Elastase and Rhamnolipid Biosynthesis Genes. *Journal of Bacteriology*, 179, pp. 5756-5767.

Retallack, D. M., Jin, H. and Chew, L. (2012). Reliable Protein Production in a *Pseudomonas fluorescens* Expression System. *Protein Expression and Purification,* 81, pp. 157-165.

Retallack, D. M., Schneider, J. C., Mitchell, J., Chew, L. and Liu, H. (2007). Transport of Heterologous Proteins to the Periplasmic Space of *Pseudomonas fluorescens* Using a Variety of Native Signal Sequences. *Biotechnology Letter,* 29, pp. 1483-1491.

Rothmel, R. K., Chakrabarty, A. M., Berry, A. and Darzins, A. (1991). Genetic Systems in *Pseudomonas. Methods in Enzymology*, 204, pp. 485-515.

Scholz, P. H., Wittman-Liebold, B., Ashman, K., Bagdasarian, M. and Scherzinger, E. (1989). Complete Nucleotide Sequence and Gene Organization of the Broad-host-range Plasmid RSF1010. *Gene*, 75, pp. 271-288.

Schumann, W. (2008). *Escherichia coli* Cloning and Expression Vectors. In: Lipps, G., ed., *Plasmids Current Research and Future Trends*, Norfolk: Caister Academic Press, pp. 1-25.

Schweizer, H. P. (1992). Allelic Exchange in *Pseudomonas aeruginosa* using Novel ColE1-type Vectors and a Family of Cassettes Containing a portable *oriT* and the Counterselectable *Bacillus subtilis sacB* Marker. *Molecular Microbiology*, 6, pp. 1195-1204.

Schweizer, H. P. (1993). Small Broad-host-range Gentamycin Resistant Gene Cassettes for Site-specific Insertion and Deletion Mutagenesis. *Biotechniques*, 15, pp. 831-834.

Schweizer, H. P. (2001). Vectors to Express Foreign Genes and Techniques to Monitor Gene Expression of Pseudomonads. *Current Opinion in Biotechnology*, 12, pp. 439-445.

Schweizer, H. P. and Hoang, T. (1995). An Improved System for Gene Replacement and *xylE* Fusion Analysis in *Pseudomonas aeruginosa. Gene*, 158, pp. 15-22.

Studier, F. W. and Moffatt, B. A. (1986). Use of Bacteriophage T7 RNA Polymerase to Direct Selective High-level Expression of Cloned Genes. *Journal of Molecular Biology*, 189, pp. 113-130.

Suh, S., Silo-Suh, L. A. and Ohman, D. E. (2004). Development of Tools for the Genetic Manipulation of *Pseudomonas aeruginosa*. *Journal of Microbiological Methods*, 58, pp. 203-212.

Thomas, C. M. (1981). Molecular Genetics of Broad Host Range Plasmid RK2. *Plasmid*, 5, pp. 10-19.

Tosi, T., Estrozi, L. F., Job, V., Guilvout, I., Pugsley, A. P., Schoehn, G. and Dessen, A. (2014). Structural Similarity of Secretins from Type II and Type III Secretion Systems. *Structure,* 22, pp. 1348-1355.

Vareechon, C., Zmina, S. E., Karmakar, M., Pearlman, E. and Rietsch, A. (2017). *Pseudomonas aeruginosa* Effector ExoS Inhibits ROS Production in Human Neutrophils. *Cell Host & Microb*e, 21, pp. 611-618.

Wang, R. F. and Kushner, S. R. (1991). Construction of Versatile Low-copy-number Vectors for Cloning, Sequencing and Gene Expression in *Escherichia coli*. *Gene*, 100, pp. 195-199.

Wang, Z., Xiong, G. and Lutz, F. (1995). Site-specific Integration of the PHICTX Genome into the *Pseudomonas aeruginosa* Chromosome: Characterization of the Functional Integrase Gene Located Close to Upstream of *attP*. *Molecular & General Genetics: MGG*, 246, pp. 72-79.

Warren, J. W., Walker, J. R., Roth, J. R. and Altman, E. (2000). Construction and Characterization of a Highly Regulable Expression Vector, pLAC11, and Its Multipurpose Derivatives, pLAC22 and pLAC33. *Plasmid*, 44, pp. 138-151.

West, S. E. H. and Iglewski, B. H. (1988). Codon Usage in *Pseudomonas aeruginosa*. *Nucleic Acids Research*, 16, pp. 9323-9335.

West, S. E. H., Schweizer, H. P., Dall, C., Sample, A. K. and Runyen-Janecky, L. J. (1994). Construction of Improved *Escherichia-Pseudomonas* Shuttle Vectors Derived from pUC18/19 and the Sequence of the Origin Required for Their Replication in *Pseudomonas aeruginosa*. *Gene*, 128, pp. 81-86.

Wilton, R., Ahrendt, A. J., Shinde, S., Douglas, D. J. S., Johnson, J. L., Brennan, M. B., et al. (2018). A New Suite of Plasmid Vectors for

Fluorescence-based Imaging of Root Colonizing Pseudomonads. *Frontiers in Plant Science*, 8, pp. 2242.

Zhang, W., Lu, J., Zhang, S., Liu, L., Pang. X. and Lv, J. (2018). Development an Effective System to Expression Recombinant Protein in *E. coli* via comparison and Optimization of Signal Peptides: Expression of *Pseudomonas fluorescens* BJ-10 Thermostable Lipase as Case Study. *Microbial Cell Factories,* 17, pp. 50.

In: *Pseudomonas aeruginosa* ISBN: 978-1-53616-100-7
Editors: C. Fah Wong and H. Idris

Chapter 2

Pseudomonas aeruginosa BIOFILMS: FORMATION, SURFACE INTERACTIONS AND ANTIMICROBIAL RESISTANCE

Nur Hazlin Hazrin-Chong*[1,*], *and
***Gee Chong Ling*[2]**
[1]School of Biosciences and Biotechnology,
Faculty of Science & Technology,
Universiti Kebangsaan Malaysia, Bangi, Selangor, Malaysia
[2]School of Biotechnology and Biomolecular Sciences,
Faculty of Science,
University of New South Wales, Sydney, NSW Australia

ABSTRACT

Pseudomonas aeruginosa is a Gram-negative, rod-shaped bacterium that is synonymous in causing widespread diseases in humans. One trait

* Corresponding Author's E-mail: hazlin@ukm.edu.my.

that is prevalent amongst *P. aeruginosa* strains is their capacity to form biofilms. Biofilms are an entity of cells adhering to a surface and/or onto each other, held and surrounded by self-produced extracellular polymeric substances (EPS). This microbial edifice, consisting of cells, polysaccharides, proteins, DNA, nutrients and waste, amongst others, is more resistant against antibacterial agents and powerful in attacking host cells compared to its free-living counterparts. This chapter begins by describing *P. aeruginosa* biofilm development, both in general terms pertaining to bacterial biofilms and specific to the bacterium. Mechanisms of cell attachment and biofilm formation based on the physico-chemical, biological and molecular perspectives are highlighted in this chapter, as these factors mostly contribute to successful *P. aeruginosa* colonisation on host surfaces. The chapter progresses to the role of *P. aeruginosa* biofilms in the environment, specifically in medical settings where the bacterium is prominent in causing infections through cell host surface interactions. Lastly, antimicrobial resistance strategies by *P. aeruginosa* biofilms are discussed before concluding with a future outlook on *P. aeruginosa* biofilm research.

INTRODUCTION

For a long time, microorganisms were viewed as simple organisms that lived unicellularly without any complex multicellular interactions. However, this perception changed with the discovery that microorganisms, apart from being able to live as single cells, can attach to surfaces and form complex, multicellular units. The formation of aggregated cells embedded in polymer matrices and highly structured communities are known as 'biofilms' (Costerton et al., 1995; Costerton, 1999; Hall-Stoodley et al., 2004).

Ubiquitous in nature, biofilms can form on almost all types of surfaces provided sufficient moisture and accessible nutrients are available (Costerton et al., 1995). Cells are thought to adhere to surfaces and form biofilms as a form of protection against physical, biological or chemical attacks in the environment (Hall-Stoodley et al., 2004). Further, cells living as biofilms have the advantage of performing tasks that would otherwise be ineffective if conducted unicellularly. Many persistent and chronic bacterial infections, for example, depend on the effectiveness of the pathogens to grow as biofilms than single, free-living cells (Costerton et al., 1999).

The importance and significance of biofilms in the world of microbiology – and beyond – cannot be underestimated. Since microorganisms generally and prominently exist as biofilms in natural settings, it is undoubted that the sessile way of life is essential to their survival. It has been shown that biofilms can withstand nutrient-starved environment such as in arid limestone fracture (Castegnier et al., 2018), and that being in biofilms can give both bacteria and fungi a fitness advantage against competitors for nutrients and space (Deschaine et al., 2018; Nadell, Drescher and Foster, 2016). Further, studies have also shown bacteria gain greater protection against predation e.g., by protozoa or parasitic bacteria when in a biofilm form as compared to living planktonically (Kadouri and O'Toole, 2005; Matz et al., 2005; Matz et al., 2008; Sun, Noorian and McDougald, 2016). Understanding how a biofilm fundamentally works helps us apply this knowledge in alleviating and/or solving many of today's problems associated with biofilms.

Biofilms are widely recognised in various fields including disease and infections (Geipel, 2009; Hawser, 1994), biofouling (Shikuma and Hadfield, 2010), bioprocessing (Wu et al., 2009), microbial fuel cells (Rabaey et al., 2007; von Canstein et al., 2008), steel corrosion (Mehanna et al., 2009; Zhang et al., 2007), wastewater treatment (Gabr et al., 2009) and in natural environments such as streams and rivers (Besemer et al., 2009; Luef et al., 2009) and hydrothermal vents (Westall et al., 2001). This shows the significance of biofilms in a diverse range of environments, which are either beneficial (e.g., in maintaining the ecosystem and through biotechnological applications) or destructive (e.g., infections and diseases) to humankind.

Studies on biofilms have begun by exploring several prominent and culturable bacteria in their ability to form a sessile community. One excellent biofilm producer that has been thoroughly investigated in terms of its biofilm development and consequent impact on humans is *Pseudomonas aeruginosa. P. aeruginosa* is a Gram-negative opportunistic pathogen capable of forming extensive biofilms. In fact, *P. aeruginosa* is argued to possess the greatest number of biofilm-producing strains (Sanchez et al., 2013). The bacterium is part of the ESKAPE pathogens (comprising of *Enterococcus faecium*, *Staphylococcus aureus*, *Klebsiella pneumoniae*, *Acinetobacter baumanii*, *Pseudomonas aeruginosa* and *Enterobacter*), a group of bacteria that are highly resistant to antibiotics and other antimicrobials (Pendleton, Gorman and Gilmore, 2013).

A prominent feature of *P. aeruginosa* in the medical field is its role in infecting patients with cystic fibrosis (CF), a genetic disorder that causes severe lung damage and breath limitation over time. Approximately 80% of CF lung infections in adult patients are caused by *P. aeruginosa* (Lyczak, Cannon and Pier, 2002) and that the infection causes death if it is not controlled (Ciofu et al., 2015). Eradicating infection caused by *P. aeruginosa* in CF patients is an almost impossible task as it is highly difficult to remove the pathogen completely from the lungs, despite administering antibiotics or other antimicrobials of various types and doses (Ciofu et al., 2015). *P. aeruginosa* possesses highly adaptive resistance mechanisms to the host defense system due to its sheer ability in forming biofilms from free-living cells (Høiby et al., 2010). Being in a biofilm form therefore gives *P. aeruginosa* strong protection for it to be able to persist in the lung environment, which causes long-term detriment to the patient.

The significance of *P. aeruginosa* in causing chronic infections through biofilm formation necessitates thorough investigation in order to understand the nature of *P. aeruginosa* cell attachment and biofilm formation, and consequently develop effective tools and mechanisms to eradicate it. This chapter first delves into the current knowledge on *P. aeruginosa* biofilm development, where physico-chemical, biological and molecular factors play a role in influencing its formation. Following that is the role of *P. aeruginosa* in the natural and medical environments - the latter focusing on *P. aeruginosa* biofilms in causing infections by colonising host surfaces. The chapter then discusses several antimicrobial resistance strategies by *P. aeruginosa* biofilms before concluding with an outlook on *P. aeruginosa* biofilm research.

P. aeruginosa AS SINGLE CELLS

Since biofilms partly consist of aggregated cells adhering to each other and/or surfaces, it is useful to first look at the *P. aeruginosa* on a cellular level before describing its biofilm formation. As a single cell, *P. aeruginosa* is a rod-shaped, Gram-negative bacterium, possessing cell appendages including adhesin, pili and flagella (Figure 2.1). These appendages, also common to other bacteria, assist *P. aeruginosa* in surface attachment and subsequently, biofilm formation. The appendages are part of the biological factors that influence biofilm formation, which are discussed further into the chapter. *P. aeruginosa* cell surface morphology also determines its physico-chemical property (e.g., hydrophobicity), thereby influencing cell-surface interaction and adhesion at the initial stage of biofilm development (Bruzaud et al., 2015). Further discussion on the physico-chemical aspects of *P. aeruginosa* cell adhesion can be found in the following sub-headings.

In addition to its cellular structure, *P. aeruginosa* is known to produce a multitude of extracellular substances that assist in its biofilm formation and contribute to its pathogenicity. These molecules play a role in cell-to-cell signalling, an important aspect in biofilm development and maintenance, and/or are part of its virulence factors when invading the host cell surface. The molecular role of *P. aeruginosa* cell adhesion and biofilm formation is elaborated further in the next few sub-chapters.

P. aeruginosa BIOFILM DEVELOPMENT

General Biofilm Formation

P. aeruginosa biofilm formation follows that of the general bacterial biofilm model, which is elaborated below through the stages of biofilm formation.

Bacteria in general form biofilms through several sequential stages (Figure 2.2), namely: 1) reversible cell attachment, 2) irreversible cell adhesion, 3) microcolony formation, 4) biofilm maturation and 5) dispersal. The formation involves distinct cell-substratum interaction (monolayer biofilm) as well as cell-cell interaction (multilayer biofilm). Also, both the beginning and end of a biofilm cycle consist of individual, free-living (planktonic) cells that result in continuous colonisation of new surfaces.

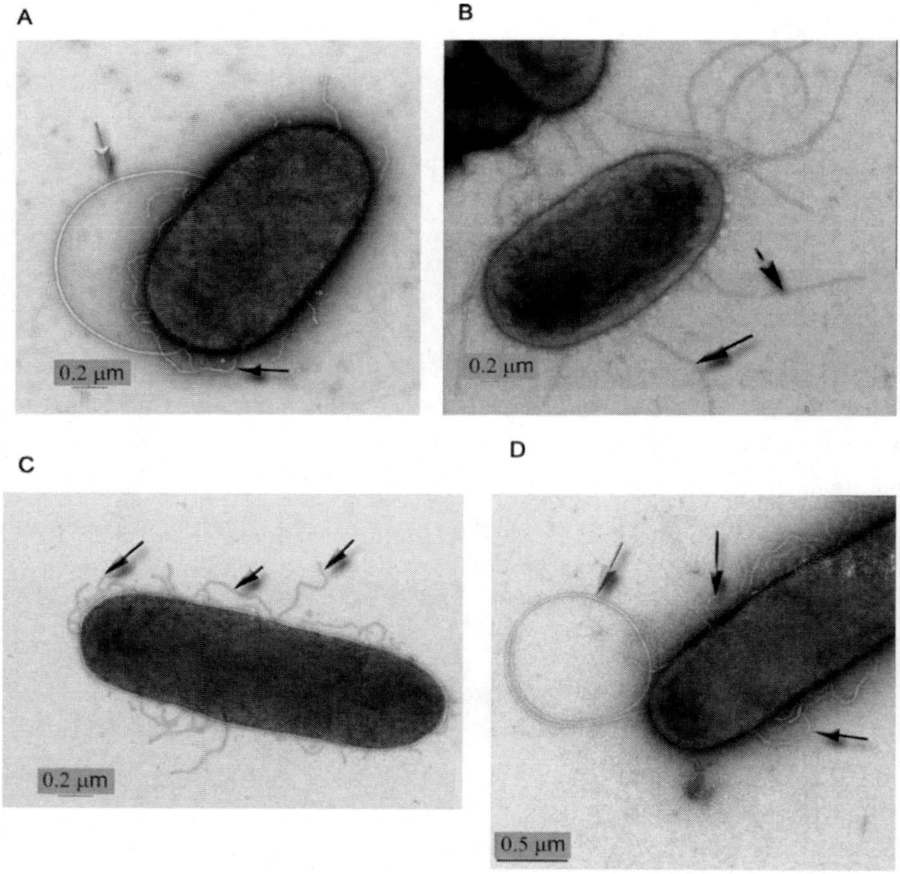

Figure 2.1. Transmission electron micrograph (TEM) of multi-drug resistant *P. aeruginosa* strains (A-D). *P. aeruginosa* is a rod-shaped Gram-negative bacterium with cell appendages including flagella (grey arrows). Unusual flagella-like structures unique to the clinical virulent isolates were also observed (Image adapted from Zaborina et al. 2008. © Creative Commons Library).

The biofilm formation model is useful as a general guide to understanding the main mechanisms of cell attachment and surface colonisation. However, differences in terms of length of each stage and colonisation behaviour are to be expected across different microbial species, communities and/or environments. Thus, the variation of biofilm development for *P. aeruginosa* will be part of the focus in this sub-chapter.

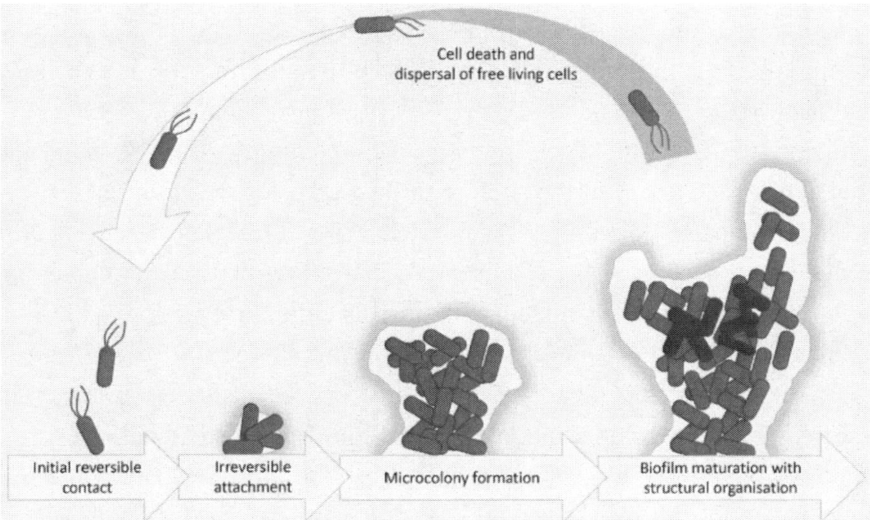

Figure 2.2. The biofilm lifecycle follows five distinct stages. It starts from: 1) initial reversible attachment of bacterial cells; 2) production of bioadhesives contributing to irreversible attachment of the bacterial cells; 3) proliferation and microcolony formation; 4) maturation of biofilm and structural organisation; and 5) cell death in the microcolonies and dispersal of single cells back into the free-living state, beginning the process over again.

The first form of contact between cells and a surface, termed reversible adhesion, begins by single cells 'loosely' attaching to the surface. This occurs mainly as a result of physico-chemical factors (discussed in the following sub-chapter), which include van der Waals forces, hydrophobicity, surface free energy, electrostatic forces, topography, surface roughness and the presence of a conditioning film (Busscher and Weerkamp, 1987; Busscher et al., 1990; Characklis, McFeters and Marshall, 1990; Fletcher and Marshall, 1982; Katsikogianni and Missirlis,

2004; Quirynen and Bollen, 1995; Van Loosdrecht et al., 1990b; Van Loosdrecht et al., 1987). These factors are only effective once the cell is close enough (within 50 nm of distance) to the surface. Thus, cells are first transported towards the surface through several different modes, including diffusion by Brownian motion, convection through liquid or air flow, and/or chemotactic activity by the microorganism (e.g., flagellar-mediated motility) (Busscher and Weerkamp, 1987; Quirynen and Bollen, 1995). These passive forces allow bacterial cells to temporarily attach to a surface substratum, and 'sample' a niche, before committing to irreversible attachment (Caiazza and O'Toole, 2004). Only favourable physico-chemical interactions between the cell and substratum would result in its initial attachment to the surface before further adhesion takes place.

Once a type of surface is found suitable for further attachment, cells use adhesive substances (i.e., 'adhesins') that allow them to permanently attach to the surface. This attachment stage is called irreversible adhesion as the cells are now committed to colonising the surface for biofilm development (Costerton et al., 1995). Three different classes of bacterial adhesins mediate the initial monolayer biofilm: (i) preformed adhesins, (ii) conditionally synthesized adhesins and (iii) specific host-related adhesins. Preformed adhesins, such as flagella, can assist cells in transient attachment to surface substrata (Lauriano et al., 2004). The presence of flagella not only supports motility but also allows bacteria to overcome long-range repulsive forces present when approaching a surface substratum (Fletcher, 1996). The acceleration of surface attachment via flagella has been identified in various bacteria, including several well-studied pathogens such as *Vibrio cholerae, Aeromonas caviae, Listeria monocytogenes, Escherichia coli, Agrobacterium tumefaciens* and *P. aeruginosa* (McClaine and Ford, 2002; Kierek and Watnick, 2003; Kirov, Castrisious and Shaw, 2004; Lemon, Higgins and Kolter, 2007; Merritt, Danhorn and Fuqua, 2007; Toutain et al., 2007). For many Gram-negative bacteria, elongated appendages such as pili also play similar roles for surface attachment and retraction movement along substrata (O'Toole and Kolter, 1998; Kang et al., 2002; Pratt and Kolter, 2002; Bechet and Blondeau, 2003; Bodenmiller, Toh and Brun, 2004; Meibom et al., 2004).

In the event of bacterial pathogenesis, it is well understood that highly specialised adhesins are synthesised to recognise and adhere to the eukaryotic host cells' receptors (Isberg, Voorhis and Falkow, 1987). The different structures and components employed by bacteria to reversibly attach onto surface substrata display diversity in surface adaptive mechanisms to achieve permanent adhesion, reflecting the many environments they inhabit. For example, the enteric pathogen, *Yersinia pseudotuberculosis*, produces a surface-associated protein called invasin, which specifically adheres to β_1 integrins on the surface intestinal epithelial cells, allowing the bacteria to enter and proliferate within the mammalian host cells (Isberg and Van Nhieu, 1995; Alrutz et al., 2001; Isberg and Barnes, 2001). Other membrane-bound proteins that assist in specific cell surface receptor adhesion include E-cadherin from *L. monocytogenes* (Mengaud et al., 1996), fibronectin from *Staphylococcus aureus* (Sinha et al., 2001), LecA and Lec B protein from *P. aeruginosa* (Tielker et al., 2005; Diggle et al., 2006) and intimin from *E. coli* (Goosney et al., 2000). These specific adhesins play an important role for the colonisation of host tissue.

Overall, the irreversible attachment stage ensures a firm anchorage of the cells onto the surface, hence preventing any potential desorption by physico-chemical forces (e.g., shear force) that are more influential in the first stage i.e., reversible attachment. At this point, cells that are attached to the surface form a monolayer of biofilm, which serves as the basis for further attachment and early biofilm growth.

Subsequent biofilm development involves microcolony formation, which consists of small clusters of cells adjoining to each other forming a multilayer biofilm. The microcolonies are developed via several ways, including multiplication (growth) of the originally adhered cells, attachment of new cells to the established monolayer and/or co-adhesion of cells in mixed-species population (Hall-Stoodley and Stoodley, 2002; Quirynen and Bollen, 1995; Stoodley et al., 2002; Tolker-Nielsen et al., 2000). In an aqueous environment, some cells mobilise themselves on the surface to co-aggregate with other cells. Different motility behaviours possessed by bacteria (e.g., type IV pili in *P. aeruginosa*, type I pili in *E.*

coli) allow the expansion of the microcolonies into different shapes and sizes depending on the species involved (Jackson et al., 2002; Klausen et al., 2003b; O'Toole and Kolter, 1998). The varying attachment strategy combinations result in different aggregation patterns for surface associated cells.

The mechanisms used for the formation of microcolonies vary depending on the bacterial cells involved, surface properties, as well as other environmental physico-chemical properties (Hall-Stoodley and Stoodley, 2002). Attached cells, for example, can also expand into microcolonies by clonal growth through cell division on surface substrata, spreading outward and upward to form a multilayer biofilm (Tolker-Nielsen et al., 2000). This growth mechanism has been observed in *Mycobacterium fortuitum* (Hall-Stoodley and Stoodley, 2005) and *P. aeruginosa* (Klausen et al., 2003b) biofilms. Clonal growth of *P. aeruginosa* involves migration of cells via twitching motility to form a dynamic flat biofilm (Klausen et al., 2003a).

Microcolony expansion progresses into a mature biofilm, which consists of cells encased in a highly structured and complex microbial 'edifice'. Typical structural characteristics of mature biofilms range from a thick, homogeneous mat of cells, to complex channels and pillars, which are thought to help transport nutrient and oxygen to the tightly packed cells and efficiently remove waste material (Costerton et al., 1995; Davey and O'Toole, 2000; Tolker-Nielsen et al., 2000). Mature biofilms not only consist of cells, but also carbohydrates, exopolysaccharides, proteins, extracellular DNA (eDNA), RNA, cell waste, ions and other matrix-soluble substances, all of which are encased in a polymer matrix called exopolymeric substances (EPS) (Sutherland, 2001a; Sutherland, 2001b). Many of the biofilm components (e.g., exopolysaccharides and eDNA) play a vital role in ensuring biofilm structural integrity and stability (Danese, Pratt and Kolter, 2000; Karatan and Watnick, 2009; Whitchurch et al., 2002), thus protecting it, to a certain extent, against various chemical, biological and physical threats (Liu and Tay, 2002; Mah and O'Toole, 2001; Matz et al., 2005). Further, cell-to-cell signalling (i.e., quorum sensing) occurs actively at this stage, in which cells produce signalling molecules (e.g., N-acyl homoserine lactone) that regulate a diverse range of physiological needs (e.g., motility, metabolite production, stress response) to ensure a fully functional biofilm.

A prominent feature in mature biofilms is the production of EPS, which form a major component of the biofilm matrix and are required for surface attachment and biofilm formation. One of the main constituents of EPS is the different types of polysaccharides used in the construction of biofilm matrices. The most commonly studied example is the polymer of β-1,6-*N*-acetyl-D-glucosamine (PGA), used by *Staphylococcus epidermidis*, *S. aureus*, *Yersinia pestis*, and *Bordetella spp*. (Itoh et al., 2005; Itoh et al., 2008; Izano et al., 2008; Choi et al., 2009; Gening et al. 2010). Negative PGA *E. coli* mutants failed to attach even after prolonged incubation. Further, chemical treatment that disrupted PGA resulted in dispersal of single biofilm cells, implicating the polymer in cell-cell adhesion as well as cell-surface interactions (Wang, Preston and Romeo, 2004).

Given the important role exopolysaccharides play in biofilm formation, it is not surprising that some bacteria have developed the capacity to produce multiple types of these matrix polymers. *P. aeruginosa*, for example, not only produces PGA but also overproduces alginate, a polymer of β-1,4-linked mannuronic acid and guluronic acid, in colonisation of the lungs of cystic fibrosis (CF) patients (Evans and Linker 1973; Govan and Deretic, 1996). This alginate overproduction gives rise to a mucoid phenotype, increased persistence and antibiotic resistance of biofilms in the lungs of the CF patients (Lam et al., 1980; Hentzer et al., 2001).

Non-mucoid *P. aeruginosa* strains have also been identified. In these strains, matrix formation is achieved via expression of the *pel* and *psl* genes. The former result in pellicle formation at the air-medium interface, while the latter facilitate attachment of lung epithelial cells by producing mucin to coat the surface (Friedman and Kolter, 2003, Friedman and Kolter, 2004; Ma et al., 2006). The *pel* and *psl* gene loci are responsible for the production of glucose-rich components, and mannose and galactose components of the matrix respectively (Matsukawa and Greenberg, 2004).

Biofilm matrix proteins also include pili and fimbriae. *E. coli*, for example, produces curli fimbriae to mediate contact with the substratum as well as other cells during biofilm formation (Prigent-Combaret et al., 2001; Gerstel et al., 2004). *P. aeruginosa* mutants without the ability to synthesise CupA fimbriae form pellicles that are easily disrupted (Friedman and Kolter, 2004). Another important component of the biofilm matrix is extracellular DNA (eDNA) (Whitchurch et al., 2002). Mature *P. aeruginosa* biofilms dissolve after DNase treatment, indicating the structural importance of eDNA within the matrix (Nemoto et al., 2003). It has been speculated that eDNA acts as a grid for the movement of bacteria by means of type IV pili binding (Van Schaik et al., 2005).

Biofilm formation ends at a stage called 'dispersal', where cells inside the biofilm are released into the environment as planktonic cells. Various factors can cause biofilm dispersal, which includes toxicity, limited oxygen and nutrient availability and increased shear stress (Barraud et al., 2006; Purevdorj, Costerton and Stoodley, 2002; Schleheck et al., 2009; Thormann et al., 2005). Bacterial cells growing within the innermost layers will suffer from these stresses, resulting in an unfavourable microenvironment. The dispersal is induced by different mechanisms by the biofilm such as cell signal-mediated inhibition of further biofilm growth or enzyme production that degrades biofilm components (Boles and Horswill, 2008; Hammer and Bassler, 2003; refer to McDougald et al., 2011 for a comprehensive review on biofilm dispersal). The dispersed cells would then search for other surfaces to colonise and begin a new biofilm formation cycle.

ROLE OF PHYSICO-CHEMICAL ADHESION ON BIOFILM FORMATION

Amongst all stages in biofilm formation, a key-determining factor to surface colonisation is the initial attachment of cells, which are primarily governed by physico-chemical factors. Cells including those of *P.*

aeruginosa need to overcome various physico-chemical obstacles in order to reach a surface, where only then they can fully implement biological mechanisms for further adhesion. Thus, it is important to consider the non-biological influences apart from the biological and molecular roles in determining the initial cell attachment from which biofilms can be formed.

Established physico-chemical theories have been used to describe cell adhesion, which hold prominence in determining the initial cell attachment to surfaces (Bos, Mei and Busscher, 1997; Bos, van der Mei and Busscher, 1999; Busscher and Weerkamp, 1987; Busscher, Sjollema and van Der Mei 1990; Van Loosdrecht et al., 1990a; Van Loosdrecht and Zehnder, 1990). It is important to note that, as with all theories and models that describe biofilm formation, the degree of influence of these factors on cell attachment are varied across different microorganisms and environments. However, the following models are widely accepted and validated as a powerful tool in understanding the physico-chemical mechanisms of cell adhesion.

To reach a particular surface, cells must overcome several energy barriers in the water or air medium that are governed by various physico-chemical elements, depending on the distance between the cell and substratum. Movement of cells at a distance of > 50 nm away from the surface are mainly governed by Brownian motion, air or water convection, or flagellar activity (Busscher and Weerkamp 1987). Within 50 nm cells must overcome two more barriers, termed 'secondary minimum' and 'primary minimum' that are determined by long-range (< 50 nm) and short-range (< 2 nm) interactions.

In accordance to the 'DLVO' theory (i.e., 'Derjaguin, Landau, Verwey and Overbeek'), long-range interactions mainly comprise of Lifshitz-Van der Waals, G_A, and electrostatic forces, G_E. These forces make up the total Gibbs energy of interaction, G_{TOT}, where a negative value would result in favourable long-range interactions. Three Van der Waals forces have been identified, i.e., induced dipole-induced dipole (London dispersion), induced dipole-dipole (Debye interaction) and dipole-dipole (Keesom interaction) that would occur between two surfaces. These forces are usually favourable between surfaces.

The electrostatic interaction, however, is largely determined by the surface charge and ionic strength of the medium. Opposite charge of two surfaces would cause favourable electrostatic interaction, whilst surfaces with the same charge would repel each other. In nature, microbial cells and surfaces are predominantly negatively charged (Quirynen and Bollen, 1995), therefore on this basis repulsion is expected. However, the distance between the cells and surface is greatly minimised at high ionic strength of the medium, hence weakening the repulsion. Through favourable van der Waals interactions, and at a high ionic strength, the total Gibbs energy of (long-range) interaction, G_{TOT} is reduced, thus enabling the cell to reach the primary minimum.

Once in primary minimum range, short-range forces such as hydrophobicity, hydrogen bonding, steric interaction and bridging interaction are present to determine the adhesion strength between the cell and the substratum surfaces (Quirynen and Bollen 1995). In an aqueous medium, a water barrier between two interacting surfaces exists and must be removed for direct contact to occur. This water barrier can be easily removed by strong hydrophobic interactions, which is determined by the hydrophobicity of cell and substratum surfaces (Boks et al., 2008). Higher cell and/or substratum surface hydrophobicity would generally result in stronger hydrophobic interactions. The barrier can also be overcome by cell appendages (e.g., pili, fimbriae) that act as 'bridges' for cell attachment to occur (Boks et al., 2008).

A favourable adhesion between the cell and substratum is given when ΔG_{Adh} is negative, whereas positive ΔG_{Adh} would result in unlikely adhesion based on thermodynamic principles. Under several different models, the interfacial energies for the above equation can be derived through experimental work on contact angle measurements based on the Young equation (Young, 1805). One of the most frequently used models is the van Oss approach (van Oss, Good and Chaudhury, 1986), which asserts that the major components of interfacial energies are Lifshitz-van der Waals and polar acid-base interactions, which are most important for thermodynamic balance (Busscher, Sjollema and van Der Mei, 1990). This model further divides the acid-base component into electron donor and electron acceptor capacities to further describe the adhesion thermodynamics (Bellon-Fontaine, Rault and van Oss, 1996). The surface thermodynamics model has been empirically tested and validated and known to be a powerful tool to predict microbial adhesion to solid substrata (Bos, van Der Mei and Busscher, 1999; Busscher et al., 1984).

Physico-Chemical Properties of *P.* a*eruginosa* Cell Surface

Given the importance of physico-chemistry in influencing cell adhesion, it is thus necessary to evaluate these properties in not only the substratum surface of interest (e.g., glass, host tissue), but also that of the cell – in this case, *P. aeruginosa*. In general, like many other bacteria, *P. aeruginosa* exhibits relatively hydrophilic and overall negative charge cell surface (Azelmad et al., 2018; Gottenbos, 2001; Mitik-Dineva et al., 2008). Nevertheless, there are differences in values between *P. aeruginosa* and those of other bacteria that warrant some mention. For example, when compared against *E. coli* and *S. aureus*, *P. aeruginosa* cell surface showed the most hydrophilic and least electronegative (Mitik-Deneva et al., 2008). This makes *P. aeruginosa* the most proficient coloniser amongst the three bacteria on glass, or other similar surfaces that are relatively hydrophilic and more negatively charged. The higher attraction resulted from a greater affinity between two hydrophilic surfaces (as opposed to one that is hydrophilic and another hydrophobic), and a greater net charge difference between *P. aeruginosa* and the substratum. Positively charged surfaces may also attract a similar if not more adhesion propensity by *P. aeruginosa* if the net charge difference is high.

Besides surface wettability and charge, surface free energy i.e., consisting of the Liftshitz-van der Waals and acid-base interactions are amongst other equally important physico-chemical factors influencing cell attachment. These properties were recently evaluated for several *P. aeruginosa* strains in a study that was conducted to measure the bacterial adhesion propensity to kitchen benchtop surfaces (Azelmad et al., 2018). Depending on the strain, *P. aeruginosa* exhibited varying levels of

Lifshitz-van der Waals surface tension; however, all tested strains showed a greater electron donating capacity and less of electron accepting property based on the acid-base interaction values obtained. Calculating the same physico-chemical values for the benchtop surfaces enabled the Gibbs free energy of adhesion (ΔG_{Adh}) between *P. aeruginosa* and the substratum to be predicted. In this case, not all *P. aeruginosa* strains were theorised to favourably attach to the benchtop surfaces due to the varying surface properties amongst the bacterial strains and substrata. Performing adhesion thermodynamics calculations gives an indication of how likely *P. aeruginosa* is to initially attach to a surface based on physico-chemical influences. This prediction, together with cell adhesion experiments, can be utilised in many industries including medical appliances or devices in which surfaces can be further manipulated to repel initial cell attachment and thus biofilm formation.

Influence of Substratum Surface Physico-Chemistry on *P. aeruginosa* Cell Attachment

The physico-chemical properties of the substratum surface also play an influential role in determining favourable or unfavourable attachment by *P. aeruginosa.* Apart from hydrophobicity and surface free energies mentioned above, another example of substratum surface physico-chemical property that should be considered is surface roughness. Mitik-Deneva et al. (2008) compared *P. aeruginosa* cell attachment between two glass surfaces, where one was smoother (through nano-scale etching) than the other. The authors found that *P. aeruginosa* cells preferred the smoother surface compared to its rougher version, noted by the significant increase in the number of attached cells onto the former. Further, *P. aeruginosa* cell dimension and EPS production were also significantly higher on the smoother surface than to its rougher counterpart, suggesting that smooth surfaces promote cell growth and biofilm formation further after initial cell attachment. Higher frequencies of cell attachment on smooth surfaces were also observed in another study conducted by Filho et al. (2018), where *P.*

aeruginosa colonised the smoothest titanium surface the most. These results imply the ease of *P. aeruginosa* cell attachment and consequently biofilm formation on devices that possess a smooth surface, e.g., in catheters, tubes or surgical implants, that may lead to nosocomial infections by *P. aeruginosa*. Thus, this calls for further innovation to be made to the substratum that takes nanoscale surface roughness into account.

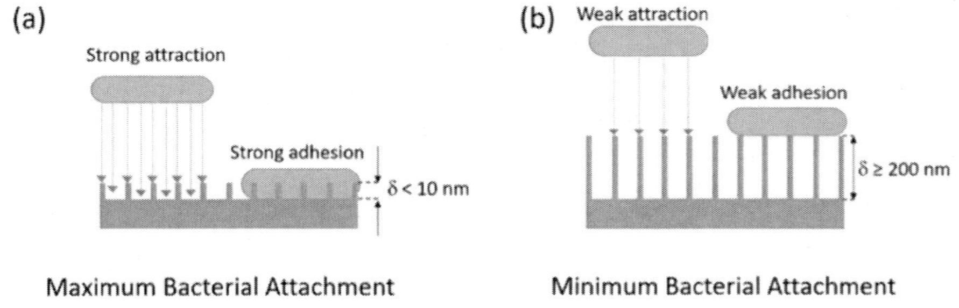

Figure 2.3. Distance (δ) between *P. aeruginosa* cells and substratum surface plays a role in promoting cell adhesion on perforated surfaces. Physico-chemical adhesion between *P. aeruginosa* cells and substratum surface is strongest when $\delta < 10$ nm (a) and weak to almost negligible as $\delta \geq 200$ nm (b). Image reprinted with permission from Lai (2018). Copyright 2018 American Chemical Society.

Interestingly, when nanoscale surface roughness is applied onto a surface to repel biofilm formation, the 'nanogratings' dimensions (height and width) should be carefully considered (Figure 2.3). An elaborate study was conducted that demonstrated this importance by utilising *P. aeruginosa* as the model bacterium (Lai, 2018). The author performed nanogratings onto a silicon substrate of varying heights and widths, observed *P. aeruginosa* cell adhesion via microscopy and performed quantitative modelling to measure physico-chemical adherence between the cell and substratum. Lai (2018) found that nanogratings which had the trenches at a certain width (i.e., smaller than the bacterial cell size by approximately an order of magnitude) and height (> 200 nm) was effective in keeping the physico-chemical adhesion between *P. aeruginosa* and substratum surface at a minimum. The decrease in width promoted cell membrane rupture by the nanoposts or capillary action that forces the cells to break down mechanically. Furthermore, trenches height that is at least 200 nm eliminates physico-chemical adhesion between the cell and substratum surface as the extended DLVO (XDLVO) forces decrease significantly with increasing distance.

Biological Factors Governing *P. aeruginosa* Biofilm Development

Biological influences play an essential part in the biofilm formation, beginning from the irreversible surface attachment to dispersal of biofilm. Microbes utilise various cell structures and substances, including extracellular polysaccharide (EPS), pili and flagella (for bacteria) and hyphae (for filamentous fungi) that significantly affect their attachment and colonisation on surfaces (Costerton et al., 1995; Jones et al., 2007; O'Toole, Kaplan and Kolter, 2000; Stoodley et al., 2002). Molecular genetic analyses showed that the production and use of these appendages involve numerous regulatory pathways, which greatly depend on environmental signals such as oxygen, nutrients, temperature, hydrodynamics and osmolarity (for reviews on molecular genetics of biofilms, see Davey and O'Toole, 2000; Molin and Tolker-Nielsen, 2003; O'Toole, Kaplan and Kolter, 2000; Stanley and Lazazzera, 2004). Further, it was shown that cell-to-cell signalling, i.e., 'quorum sensing', is essential in regulating biofilm formation through various regulatory pathways, which involve small-molecule autoinducer signalling mechanisms (Camilli and Bassler, 2006; Davies et al., 1998). These genotypic and phenotypic regulations play an integral part in both the structure and function of biofilms.

In contrast to its planktonic lifestyle, congregation of microbes during biofilm formation promotes intra- and inter-species interaction, protecting bacteria against environmental pressures such as biological predation and chemical stresses (Matz et al., 2005; Barraud et al., 2006). Formation of different biofilm structures in response to environmental conditions provides flexibility and a competitive advantage over eukaryotic organisms (Hall-Stoodley, Costerton and Stoodley, 2004). For example, biofilms grown in a laboratory flow chamber observed mushroom-shaped microcolonies and loosely shaped structural organisation (Tolker-Nielsen et al., 2000). Biofilms formed on surfaces exposed to river water exhibit microcolonies that are parallel to the ridged surface structure and direction of water flow (Battin et al., 2003, Neu & Lawrence, 2006). Similar observation was made with mixed species marine biofilms on ridged microfabricated surfaces, whereas biofilms resembling loosely protruding mushroom-shaped structure were formed on flat surfaces exposed to the same environment. Differences in biofilm structure were also found to influence microbial community composition and resistance to environmental stimuli (Ling et al., 2014). In addition to structural flexibility, biofilm formation allows the growth of bacteria at various contaminated sites that are toxic to other organisms (Singh, Paul and Jain, 2006). The cooperation among bacteria provides a range of enzymes capable of breaking down contaminants and using them as a food source (Massol-Deya et al., 1995). Biological interactions within a bacterial biofilm carry significant ecological function and are impacted by the diversity of the biofilm communities.

Bacterial biofilms contain populations or communities of cells enclosed in a matrix of extracellular polymeric substances (EPS). The matrix encapsulated cells are extremely difficult to eradicate, hence biofilms have become the source of numerous recalcitrant infections (Lewis, 2001). A report from the US National Institute of Health stated that 60% of all microbial infections are caused by biofilms (Fux et al., 2005). Biofilm-induced resistance is the main contributor to the widespread impact of these infections. The antimicrobial treatments of these infections are effective in removing the dispersed planktonic cells (Saye 2007). However, the surviving bacterial cells within biofilms can tolerate concentrations of antibiotics higher than the maximum deliverable dosage (Ceri et al., 1999; Cerca et al., 2005). As a result, widespread use of antimicrobial agents in medical, industrial and other settings contribute to increased tolerance in bacterial pathogens and the evolution of multidrug-resistant strains (Russell, 2003; Mathur, 2005).

Biofilm cells generally have a decreased growth rate compared to their planktonic counterparts (Donlan, 2001; Singh et al., 2002), thereby further reducing antibiotic efficacy as antimicrobials typically target more rapidly growing cells (Lewis, 2001). Some antibiotics, such as penicillin and ampicillin, have an absolute requirement for cell growth in order to kill and render them effective (Blumberg and Strominger, 1974; Spratt and Spratt, 1980). Other antibiotics with dual mode of actions such as cephalosporins and aminoglycosides can kill non-growing cells, but are generally more effective against actively dividing cells, since their activity mainly target protein biosynthesis (Costerton et al., 1999; Adair et al., 2002).

MOLECULAR FACTORS ENABLING A VERSATILE *P. aeruginosa* BIOFILM CAPABILITY

The versatility of *P. aeruginosa* in adapting to diverse environments is likely due to its relatively large genome, ranging from 6.22 to 6.91 Mb (Silby et al., 2011). The sequences revealed low number of unique genes to individual strains sequenced, while carrying a highly conserved core genomes of up to 90% encoding accessory genes (Mathee et al., 1999). Conservation of the core genome of *P. aeruginosa* boosts the presence of the biofilm regulatory genes, which is one of the leading factors contributing to the *P. aeruginosa* virulence ability.

It is suggested that the evolution of *P. aeruginosa* is driven by the selection pressure for environmental adaptability (Silby et al., 2011). *P. aeruginosa* is well known to be metabolically versatile as the bacterium has been isolated from nutrient-poor settings. The ability to utilise hydrocarbon and tricarboxylic acid intermediates over sugar in *P. aeruginosa* has largely enabled by the tri-carboxylate transport system identified (Stover et al., 2000; Moreno et al., 2009). Further, the number of regulatory genes in *P. aeruginosa*, amounted to nearly 10% of its genome, has also contributed to its ability to survive in more diverse and harsher environments than its counterparts (Stover et al., 2000). One such tool employed by *P. aeruginosa* to adapt to the changing environment is the two-component class (TCS) regulatory system. TCS is primarily a primitive signal transduction system comprises of a membrane bound sensor kinase that detects an extracellular signalling molecule which trigger cellular change via cytoplasmic response regulator through a cascade of enzymatic reaction. Examples of TCS system in *P. aeruginosa* include the GacA/LadS/RefS regulatory circuit (virulence), WspR-WspE (biofilm formation) and PhoP-PhoQ (antibiotic resistance) (McPhee et al., 2006; Güvener and Harwood 2007; Williams and Cámara 2009). Equipped with the arsenal of surface adaptation genes, *P. aeruginosa* is primed to habituate surface-associated lifestyle. This lifestyle of *P. aeruginosa* involves complex host-associated regulatory pathway on biotic surfaces, mostly studied on human host, as well as a diverse range of abiotic surfaces in the environment, which will be discussed in the following sections.

P. aeruginosa BIOFILM FORMATION IN THE NATURAL ENVIRONMENT – THE GOOD AND THE BAD

Biofilm formation on surfaces is highly differentiated and imparts important cellular survival traits for microorganisms in natural environments (Dworkin, 1985). *P. aeruginosa* biofilms inhabit a vast array of environmental niches including surfaces of oceanic rocks (Pohl and Schneider, 2002) and terrestrial soil (Gaylarde et al., 2004), which make them one of the key drivers of global biogeochemical cycling of key elements such as oxygen, nitrogen and carbon (Krumbein et al., 2003; Dupraz and Visscher, 2005). Moreover, the biofilm lifestyle allows bacteria to grow and stay in environments of fast-moving water such as streaming riverbeds (Battin, 2000), as well as other extreme habitats including acid mine drainage (Baker and Banfield, 2006) and deep ocean hydrothermal vents (Nichols et al., 2005). Microbial biofilm formation in the marine environment is a key stage preceding biofouling by higher organisms such as sessile invertebrates and algae.

In contrast to its planktonic mode of growth, congregation of microbes during biofilm formation promotes intra- and inter-species interaction, protecting bacteria against environmental pressures such as biological predation and chemical stresses (Matz et al., 2005b; Barraud et al., 2006). Formation of different biofilm structures in response to environmental conditions provides flexibility and a competitive advantage over eukaryotic organisms (Hall-Stoodley et al., 2004). For example, mushroom-shape biofilms are common in quiescent water, while thick flat biofilms resulted in fast flowing environments because of shear stress (Hall-Stoodley et al., 2004). In addition to structural flexibility, biofilm formation allows the growth of bacteria at various contaminated sites that are toxic to other organisms (Singh et al., 2006). The cooperation among bacteria provides a range of enzymes capable of breaking down contaminants and using them as a food source (Massol-Deyá et al., 1995).

On the other hand, the versatile *P. aeruginosa* creates negative impact when it gets into the water system (Mena and Gerba, 2009; Rice et al., 2012). The persistence of *P. aeruginosa* in water systems such as whirlpool, spas, and hot tubs has contributed to disease outbreaks, in the form of skin rashes as well as infections of the ear, eye, urinary tract and hair follicles (Salmen et al., 1983; Willcox, 2007; Yu et al., 2007; Mittal et al., 2009). The cause of these outbreaks has been linked to the increased number of *P. aeruginosa* cells in the water sample analysed, despite the presence of chlorinated disinfectants (Lutz and Lee, 2011). The elevated water temperature and presence of disinfectants were found to favour biofilm-forming bacteria such as *P. aeruginosa*, complemented with the bacterium's ability to produce proteases and lecithinases that led to inflammatory responses from the host (Highsmith et al., 1985). Further, high water temperatures have been linked to increased perspiration and desquamation, which act as a nutrition source for microbial growth (Price and Ahearn, 1988). In another study, the presence of *P. aeruginosa* in swimming pools has led to an outbreak in Britain (Tate, Mawer and

Newton, 2003), where *P. aeruginosa* biofilms were found on pool filters, carpets, shower floors and toys (Leoni et al., 1999; Uhl and Hartmann, 2005). The contamination of pathogens such as *P. aeruginosa* has also been linked to increased gastroenteritis (Dale et al., 2009). Transmission routes include epidermal contact and infection through the skin, inhalation of aerosol and direct ingestion of water (Rice et al., 2012).

Host Surface Interaction with *P. aeruginosa* Biofilms

Apart from *P. aeruginosa* being highly versatile and adaptable to a wide range of natural environments, the bacteria are also ubiquitous in many host-associated environments, including human, plants and animals (Da Silva et al., 2008, Mena and Gerba, 2009). *P. aeruginosa* biofilm formation on living host surfaces is commonly associated with the presence of environmental signals, which regulate the different stages of biofilm development (Stanley and Lazazzera, 2004). Once attached, the cells regulate production of a range of surface colonising components to assist multilayer biofilm formation, where cell-cell interactions are pivotal regulating factors. Each stage of biofilm development is tightly regulated by the presence of environmental stimuli (O'Toole and Kolter, 2002). Bacterial cells sense and respond to different environmental signals such as nutritional status and quorum sensing molecules, which regulate their growth and lifestyle to improve the chances of survival (Kjelleberg and Molin, 2002). The presence of *P. aeruginosa* biofilm in the host-associated environment is regulated through complex mechanisms and commonly resulted in many forms of infection.

P. aeruginosa Biofilm in the Medical Environment – The Infections

Apart from being fundamental to adhesion and growth on surfaces, biofilm formation promotes increased antimicrobial resistance, which then prolongs and exacerbates infections (Hall-Stoodley, Costerton and Stoodley, 2004). The formation of bacterial biofilms on medical devices such as intravenous catheters, joint prostheses, cardiac pacemakers and endotracheal tubes results in a range of device-related biofilm infections (Peters, Locci and Pulverer, 1981; Marrie, Nelligan and Costerton, 1982; Peters, Locci and Pulverer, 1982; Bauer et al., 2002). However, biofilm related infections are not restricted to abiotic surface attaching events as *P. aeruginosa* have also been known to cause host-associated infections through biotic interactions with the host, most notably human (Figure 2.4) (Von Eiff et al., 1999; Akiyama et al., 2002; Götz, 2002, Donlan and Costerton, 2002).

A common *P. aeruginosa* biofilm-related infection in humans involves mucus build-up in the lungs of cystic fibrosis (CF) patients (Gellatly and Hancock, 2013). CF is an autosomal recessive disease caused by mutation in the cystic fibrosis transmembrane conductance regulator (CFTR) gene. The mutation leads to dysfunctional electrolyte secretion and absorption, leaving CF patient with viscous respiratory mucous and impaired clearance (Lyczak, Cannon and Pier, 2002). Other associated pathogens for CF disease include *S. aureus*, and *Haemophilus influenzae* (Koch and Høiby, 1993). *P. aeruginosa* is commonly embedded within the biofilm matrix associated with sputum from CF patients and exhibits a mucoid phenotype different from *P. aeruginosa* environmental isolates. It suggests a change in gene expression resulting from the host inflammatory response. The changes include loss of flagellum, pili and type III secretion system required for host surface interaction, motility and virulence mechanism. Mucoidal variants due to mutation in the *mucA*, *mucB* or *mucD* genes gain protection from the innate immune system (Mathee et al., 1999). Another host immune response includes CF bronchial secretions, which possess proteolytic activities against fibronectin in the respiratory mucosa and has found to favour *P. aeruginosa* biofilms over *S. aureus* colonisation (Suter et al., 1988). In a healthy individual, mucin glycolipids secretion by the secretory cells on the epithelial lining will bind and trap foreign particle, assisted by the unidirectional rhythmic beating of ciliated epithelial cell for mucous clearance.

In response to host defence mechanisms, *P. aeruginosa* produces several virulence factors that contribute to respiratory pathogenesis (Figure 2.5). *P. aeruginosa* possesses surface-associated adhesins, such as flagella, responding to the receptors of the host cells. Some specific examples include adhesion of bacterial cells to the exposed asialoganglioside-binding site and the fibronectin Arg-Gly-Asp (RGD) receptors of the host epithelial cells (Saiman and Prince, 1993; Roger et al., 1999). The binding elicits a strong inflammatory response via signalling through toll-like receptor 5 (TLR5) and a caspase-1-mediated response through the Nod-like receptor, Ipaf (Miao et al., 2007). Type IV pili is a surface adhesin that contributes to the twitching mobility and biofilm formation of *P. aeruginosa* (O'Toole and Kolter, 1998; Mattick, 2002). Pili give rise to microcolonies formation through cell-cell aggregation, offering protection from the host immune system (Weitere et al., 2005). Besides surface-associated adhesins, *P. aeruginosa* possesses Type III secretion system which permits the translocation of effector proteins into the host cells. The secretion of effector protein such as ExoU, which is a phospholipase, could cause necrosis in the host epithelial cells (Hauser, 2009).

In addition, quorum-sensing (QS) signalling and anaerobic growth are important features for *P. aeruginosa* biofilms in CF patients (Xu et al., 1998; Singh et al., 2000). The production of QS molecule 3-oxo-C_{12}-homoserine lactone (HSL) by *P. aeruginosa* was found to induce the expression of host associated inflammatory cytokines and chemokines, which prolonged inflammation and lung diseases (Smith et al., 2002). Further, the dehydrated airway surface in CF lungs cause neutrophil, attracted by the increased cytokines, to be trapped at a localised site and induce neutrophil necrosis rather than apoptosis (Hayes et al., 2011).

QS leads to maturation of biofilm formation, allowing cells to be encased in extracellular polymeric substances (EPS). The biofilm lifestyle imparts protection for *P. aeruginosa* from stressors such as the host immune system and antibiotic medication, thus contributing to ongoing difficulties in treatment, observed by Drenkard and Ausubel (2002). The study showed the presence of airway mucus causing hypoxic environment that led to the production of alginate, causing stronger adhesion by *P.*

aeruginosa cells. Vigorous biofilms under anaerobic condition have been associated with the overexpression of outer membrane porin OprF by *P. aeruginosa* due to the build-up of toxic nitrogen metabolites in CF patients (Yoon et al., 2002). The shift between sessile and motile states in biofilm regulation allows the cells to disperse and colonise onto new surfaces to initiate another cycle of pathogenic event.

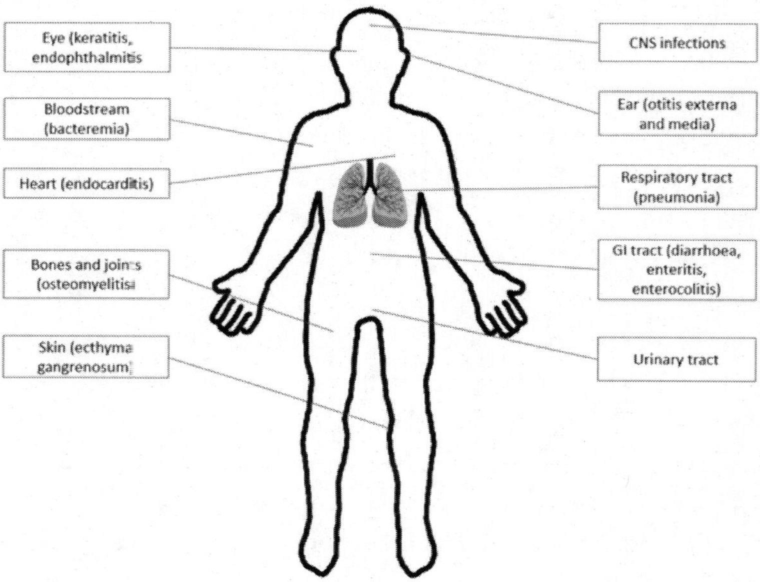

Figure 2.4. A multitude of *P. aeruginosa* infections on human host.

P. aeruginosa also secretes proteases to degrade immunoglobulin and fibrin, which disrupts host's epithelial tight junction and thus contributes to respiratory infections (Kipnis et al., 2006). Alkaline protease is a type 1 zinc metalloprotease that is responsible for the degradation of host complement protein, interfering with flagellin signalling through host TLR5 to avoid immune detection (Laarman et al., 2012). Elastases such as LasB produced by *P. aeruginosa* contributes to the degradation of the opsonising lung surfactant protein, which protects against phagocytosis (Kuang et al., 2011). Protease IV produced by *P. aeruginosa* was found to be involved in corneal infection by degrading immunoglobulin and fibrinogen, observed in a rabbit model study (Engel et al., 1998). *P.*

aeruginosa cells rely on lipopolysaccharide on its outer membrane to confer antigenicity, respond to inflammatory responses and mediate interaction with antibiotics (King et al., 2009). Lipid A is an atypical glycolipid that anchors the lipopolysaccharide into the outer membrane. The end of Lipid A could bind to host cell co-receptors and activate a signalling pathway cascade which triggers the production of pro-inflammatory cytokines and chemokines, leading to endotoxic shock (Teghanemt et al., 2005). The expression of structurally and serologically distinct lipopolysaccharide with O-antigen has been found to confer advantage in evasion of host's adaptive immune response (King et al., 2009).

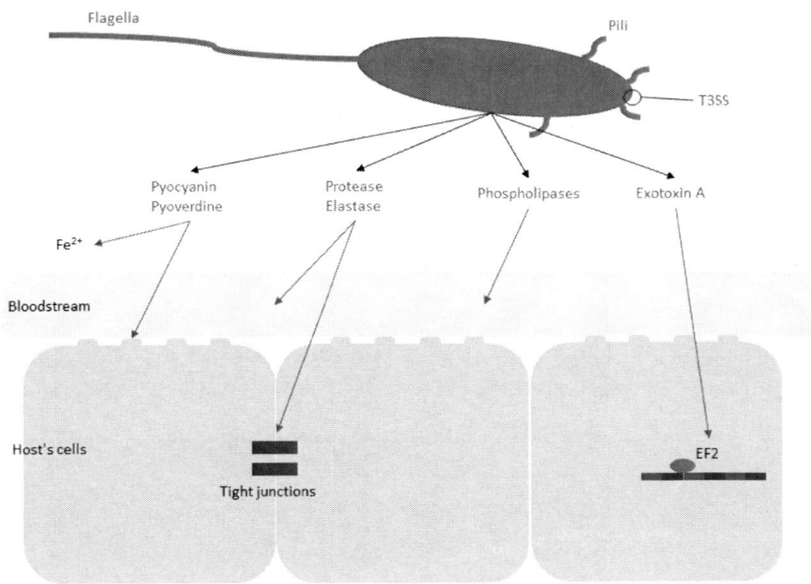

Figure 2.5. A multitude of virulence factors produced by *P. aeruginosa* in response to the host immune system. *P. aeruginosa* produces surface associated adhesins (e.g., flagella and type IV pili) which are capable to bind to host's epithelial gangliosides. The contact triggers T3SS system where cytotoxins are injected into the host cell. The virulence factors (e.g., lipases, proteases, pyocyanin - depicted in the figure) cause different degrees of impact on the host cell including the degradation of host cell membrane by targeting the tight junction and lipid surfactant. In addition, pyoverdine produced by *P. aeruginosa* enables iron scavenging from the environment, thus enhancing pathogenicity of the bacterium (image adapted from Gellatly and Hancock, 2013).

Apart from proteases, *P. aeruginosa* is known to produce other host associated virulence factors. This includes the production of Exotoxin A, which inhibits host elongation factor 2, thereby stopping protein synthesis and causes cell death (Yates et al., 2005). Further, lipases and phospholipases which break down lipid surfactants the of host's cell membrane enable cell lysis (Kipnis et al., 2006). Production of pyocyanin by *P. aeruginosa* has found to induce neutrophil apoptosis through increased oxidative stress (Lau et al., 2004; Horke et al., 2010). The complex interaction between *P. aeruginosa* and host-associated or environmental cues has resulted in unique responses and changes in the regulation of biofilm formation and currently attracts much attention in clinical research.

Other than CF patients, the presence of *P. aeruginosa* in the host system also leads to other medical infections such as septicaemia, endocarditis, osteomyelitis, urinary tract infections, gastrointestinal tract infections, burn patients, skin infections, ear infections, eye infections and meningitis. Majority of these infections occurred following an immunocompromised event that allow penetration of the skin cells. The host-associated virulence factors involved are like those that have been mentioned in this section. The adaptability of *P. aeruginosa* on surfaces in the environment is likely due to the presence of its diverse genome contributing to its pathogenicity. This also contributes to its enhanced resistance to environmental challenges, including the presence of antimicrobials.

ANTIMICROBIAL RESISTANCE STRATEGIES OF *P. aeruginosa* AND ITS BIOFILM

P. aeruginosa employs numerous strategies to resist environmental challenges, which include the host's immune response to eradicate infections caused by *P. aeruginosa* biofilms. Examples of biofilm resistance include restricted penetration of antimicrobial agents into the biofilms, expression of resistance genes and decreased growth rate in biofilm populations (Gilbert, Das and Foley, 1997; Costerton, Stewart and Greenberg, 1999; Nield-Gehrig, 2005). In general, resistance by *P. aeruginosa* can be categorised into intrinsic, acquired and adaptive types as discussed below.

Intrinsic Resistance

P. aeruginosa possesses several intrinsic antimicrobial resistance mechanisms, including selective permeability of outer membrane complexes, inducible β-lactamase production, and the presence of multidrug resistance (MDR) efflux pumps to exclude harmful molecules (De Kievit et al., 2001; Lewis, 2008). The outer membrane of *P. aeruginosa* was found to be less permeable than *E. coli* by possessing more of the smaller channel porins formed by OprB and OprD compared to the larger ones formed by OprF (Livermore, 2002; Fito-Boncompte et al., 2011). The observation contributes to the restrictive uptake of hydrophilic molecules including β-lactam antibiotics. The efflux pump formed by MexAB-OprM and MexXY-OprM can effectively excrete antimicrobial compounds including β-lactam, chloramphenicol, fluoroquinolones, tetracycline, and many others (Li, Nikaido and Poole, 1995; Pamp et al., 2008). In the event where β-lactam enters the cell, an inducible β-lactamase enzyme regulated by AmpC (located in the periplasm) will be produced, which leads to hydrolysation of the β-lactams (Giwercman et al., 1991, Mulligan et al., 1993, Bisson et al., 2002). *P. aeruginosa* biofilms also displayed an induction of β-lactamase production during piperacillin and ceftazidime treatment (Ciofu et al., 2009). The regulation of the inducible enzyme is tightly linked to the effectiveness of the outer membrane and efflux pump to restrict β-lactams from entering the cell (Jacoby, 2009).

Acquired and Adaptive Resistance

Genome plasticity has enabled *P. aeruginosa* to acquire genetic elements that confer antimicrobial resistance through lateral gene transfer processes such as transformation, transduction and conjugation. These processes are termed 'acquired resistance'. Genetic variation or mutational events may result in the *P. aeruginosa* strain with enhanced resistance. A mutation in mexZ, a natural suppressor of mexXY, has led to the overexpression of mexXY efflux pump (Vogne et al., 2004). Another mutation in DNA gyrase was found to reduce the binding affinity of fluroquinolone, causing an increased resistance to the drug by *P. aeruginosa* (Nakajima et al., 2002).

Rather than relying on the environmentally acquired elements, *P. aeruginosa* is also capable to respond to external stressors through adaptive changes in its physiology and genetic regulation. For instance, a diffusion barrier is formed by the EPS layer, protecting biofilm cells against both large macromolecules and smaller positively charged aminoglycosides (Shigeta et al., 1997; Ishida et al., 1998). The EPS diffusion barrier causes some antimicrobials to reach biofilm cells at a much slower rate, a process that is synergistically regulated with the expression of resistance enzymes such as β-lactamase, which destroy the activity of many antibiotics (Powers et al., 1999). Furthermore, *P. aeruginosa* also produces catalase, which protects the cells from hydrogen peroxide and resulting oxidative stress from the antimicrobial agent (Elkins et al., 1999; Fux et al., 2005). *P. aeruginosa* also responds to environmental stressors such as pH, heat shock, DNA stress and many others via biofilm formation. A well-studied adaptive mechanism of *P. aeruginosa* involves the upregulation of PhoPQ and PmrAB, altering permeability of the outer membrane, resulting in tolerance to polymyxins and other cationic antimicrobial peptides (Barrow and Kwon, 2009). Resistance conferred by bacterial biofilms including those of *P. aeruginosa* is thoroughly discussed in a number of reviews (Hancock and Speert, 2000; Gellatly and Hancock, 2013; Jolivet-Gougeon and Bonnaure-Mallet, 2014; Taylor, Yeung and Hancock, 2014).

CONCLUSION

Today, knowledge on *P. aeruginosa* cell attachment and biofilm formation has been established based on the extensive research conducted on the bacterium. These studies highlight *P. aeruginosa* as a model organism for fundamental studies on cell adhesion and biofilm formation, and as a pathogen causing chronic and persistent infections in humans. Information on the bacterium regarding cell surface physico-chemical characteristics and its phenotypic and genotypic properties relating to cell adhesion are now widely available. Further, the mechanisms by which *P. aeruginosa* forms biofilms in general and specific conditions have also been thoroughly investigated, either through the manipulation of the external environment or the genes that confer certain cell adhesion characteristics. Mechanisms of antimicrobial resistance by *P. aeruginosa* are also well known that aid in our understanding of the complexity surrounding biofilm-associated *P. aeruginosa* infections. Strategies to eliminate *P. aeruginosa* biofilms have also been heavily studied which provide further basis and understanding on how *P. aeruginosa* biofilms behave under various conditions.

Despite the available findings, gaps of knowledge related to *P. aeruginosa* biofilms are still wide. A high proportion of studies on *P. aeruginosa* biofilms have been conducted in laboratory settings, which may not reflect the true conditions and thus behaviour of the bacterium in their natural environment. Conducting *P. aeruginosa* biofilm studies by simulating the conditions in the environment (e.g., combination of low carbon availability with mixed microbial community) should provide further insights into its biofilm behaviour, although notably such experimental design is more challenging to construct than the standard laboratory-based culturing procedures. Furthermore, studies on *P. aeruginosa* biofilms in causing other chronic and persistent infections, e.g., septicaemia, endocarditis and urinary tract infections, have not been conducted as extensive as those conducted on CF infections. Although research on *P. aeruginosa*-related CF infections may provide the basis for other diseases caused by *P. aeruginosa*, more disease-specific studies need to be conducted as the exact mechanisms of cell attachment and resistance may be different.

Because it is now widely known that *P. aeruginosa* biofilms are notoriously resistant to many antimicrobials including antibiotics, it is expected that more studies will be focused in using newly developed tools and mechanisms to eradicate the sessile bacterium. Synthetic and natural peptide polymers, e.g., lipopolypeptide and the recent 'structurally nano-engineered antimicrobial peptide polymers' (SNAPPS) are such examples. The polymer showed high effectiveness against and long-term non-resistance by *P. aeruginosa*, although efficacy for many peptide polymers are still unestablished for biofilms compared to planktonic cells (Hincapié, Giraldo and Orduz, 2018; Lam et al., 2016; Schneider-Futschik et al., 2018). Further, research on micro- and nanoscale alterations to the substratum surface e.g., on medical devices that are prone to *P. aeruginosa* biofilm infections are also receiving an increasing amount of attention and expected to shed more light on providing effective design principles to manufacture antibiofilm surfaces (Ling et al., 2014; Lai, 2018).

Based on current research trends, it is evident that studies on *P. aeruginosa* and its biofilm remain essential even until today. This is due to the bacterium's prevalence and relevance in our environment and its significant detrimental impact towards human lives due its pathogenicity. Also, because *P. aeruginosa* is used as model bacterium in many fundamental biofilm studies, the bacterium acts as knowledge feeder in providing insights into biofilm development and behaviour across all biofilm-producing bacteria. The knowledge generated through *P. aeruginosa* biofilm studies also serves an important purpose, that is, to further develop solutions to persistent problems associated with biofilms by both *P.* aeruginosa and its counterparts.

References

Akiyama, H., Hamada, T., Huh, W. K., Yamasaki, O., Oono, T., Fujimoto, W. and Iwatsuki, K. (2003). Confocal Laser Scanning Microscopic

Observation of Glycocalyx Production by *Staphylococcus aureus* in Skin Lesions of Bullous Impetigo, Atopic Dermatitis and Pemphigus Foliaceus. *British Journal of Dermatology*, 148(3), pp. 526-532.

Alrutz, M. A., Srivastava, A., Wong, K. W., D'Souza-Schorey, C., Tang, M., Ch'Ng, L. E., Snapper, S. B. and Isberg, R. R. (2001). Efficient Uptake of *Yersinia pseudotuberculosis* via Integrin Receptors Involves a Rac1–Arp 2/3 Pathway that Bypasses N-WASP Function. *Molecular Microbiology*, 42(3), pp. 689-703.

Azelmad, K., Hamadi, F., Mimouni, R., El boulani, A., Amzil, K. and Latrache, H. (2018). Physicochemical Characterization of *Pseudomonas aeruginosa* Isolated from Catering Substratum Surface and Investigation of Their Theoretical Adhesion. *Surfaces and Interfaces*, 12, pp. 26-30.

Barraud, N., Hassett, D. J., Hwang, S. H., Rice, S. A., Kjelleberg, S. and Webb, J. S. (2006). Involvement of Nitric Oxide in Biofilm Dispersal of *Pseudomonas aeruginosa*. *Journal of Bacteriology*, 188(21), pp. 7344-7353.

Barrow, K. and Kwon, D. H. (2009). Alterations in Two-Component Regulatory Systems of *phoPQ* and *pmrAB* are Associated with Polymyxin B Resistance in Clinical Isolates of *Pseudomonas aeruginosa*. *Antimicrobial Agents and Chemotherapy*, 53(12), pp. 5150-5154.

Battin, T. J., Kaplan, L. A., Newbold, J. D., Cheng, X., & Hansen, C. (2003). Effects of Current Velocity on the Nascent Architecture of Stream Microbial Biofilms. *Applied Environmental Microbiology*, 69(9), 5443-5452.

Bauer, T. T., Torres, A., Ferrer, R., Heyer, C. M., Schultze-Werninghaus, C. and Rasche, K. (2002). Biofilm Formation in Endotracheal Tubes: Association between Pneumonia and the Persistence of Pathogens. *Monaldi Archives for Chest Disease*, 57(1), pp. 84-87.

Bazaka, K., Jacob, M. V., Crawford, R. J. and Ivanova, E. P. (2012). Efficient Surface Modification of Biomaterial to Prevent Biofilm Formation and the Attachment of Microorganisms. *Applied Microbiology and Biotechnology*, 95(2), pp. 299-311.

Bechet, M. and Blondeau, R. (2003). Factors Associated with The Adherence and Biofilm Formation by *Aeromonas caviae* on Glass Surfaces. *Journal of Applied Microbiology*, 94(6), pp. 1072-1078.

Bellon-Fontaine, M. N., Rault, J. and van Oss, C. J. (1996). Microbial Adhesion to Solvents: A Novel Method to Determine the Electron-Donor/Electron-Acceptor or Lewis Acid-Base Properties of Microbial Cells. *Colloids and Surfaces B: Biointerfaces*, 7(1-2), pp. 47-53.

Besemer, K., Hödl, I., Singer, G. and Battin, T. J. (2009). Architectural Differentiation Reflects Bacterial Community Structure in Stream Biofilms. *The ISME Journal*, 3(11), p.1318.

Bisson, G., Fishman, N. O., Patel, J. B., Edelstein, P. H. and Lautenbach, E. (2002). Extended-Spectrum Β-Lactamase–Producing *Escherichia coli* and *Klebsiella species*: Risk Factors for Colonization and Impact of Antimicrobial Formulary Interventions on Colonization Prevalence. *Infection Control and Hospital Epidemiology*, 23(5), pp. 254-260.

Bodenmiller, D., Toh, E. and Brun, Y. V. (2004). Development of Surface Adhesion in *Caulobacter crescentus*. *Journal of Bacteriology*, 186(5), pp. 1438-1447.

Boks, N. P., Norde, W., van der Mei, H. C. and Busscher, H. J. (2008). Forces Involved in Bacterial Adhesion to Hydrophilic and Hydrophobic Surfaces. *Microbiology*, 154(10), pp. 3122-3133.

Boles, B. R. and Horswill, A. R. (2008). *Agr*-mediated Dispersal of *Staphylococcus aureus* Biofilms. *PLoS Pathogens*, 4(4), pp. e1000052.

Bos, R., Mei, H. C. and Busscher, H. J. (1997). Physico-chemical Interactions in Initial Microbial Adhesion and Relevance for Biofilm Formation. *FEMS Microbiology Reviews* 11(1), pp. 24–32.

Bos, R., van der Mei, H. C. and Busscher, H. J. (1999). Physico-Chemistry of Initial Microbial Adhesive Interactions – Its Mechanisms and Methods for Study. *FEMS Microbiology Reviews*, 23(2), pp. 179-230.

Bruzaud, J., Tarrade, J., Coudreuse, A., Canette, A., Herry, J., Taffin de Givenchy, E., Darmanin, T., Guittard, F., Guilbaud, M. and Bellon-Fontaine, M. (2015). Flagella but not type IV pili are involved in the initial adhesion of *Pseudomonas aeruginosa* PAO1 to hydrophobic or superhydrophobic surfaces. *Colloids and Surfaces B: Biointerfaces*, 131, pp. 59-66.

Busscher, H. J. (1990). Relative Importance of Surface Free Energy as a Measure of Hydrophobicity in Bacterial Adhesion to Solid Surface. *Microbial Cell Surface Hydrophobicity*, pp. 335-359.

Busscher, H. J. and Weerkamp, A. H. (1987). Specific and Non-Specific Interactions in Bacterial Adhesion to Solid Substrata. *FEMS Microbiology Letters*, 46(2), pp. 165-173.

Busscher, H. J., Sjollema, J. and van der Mei, H. C. (1990). Relative Importance of Surface Free Energy as a Measure of Hydrophobicity in Bacterial Adhesion to Solid Surfaces. In: R. J. Doyle and M. Rosenberg, ed., *Microbial Cell Surface Hydrophobicity*, Washington: American Society for Microbiology, pp. 335-359.

Busscher, H. J., Weerkamp, A. H., van der Mei, H. C., Van Pelt, A. W., de Jong, H. P. and Arends, J. (1984). Measurement of the Surface Free Energy of Bacterial Cell Surfaces and Its Relevance for Adhesion. *Applied and Environmental Microbiology*, 48(5), pp. 980-983.

Caiazza, N. C. and O'Toole, G. A. (2004). *SadB* is required for the Transition from Reversible to Irreversible Attachment during Biofilm Formation by *Pseudomonas aeruginosa* PA14. *Journal of Bacteriology*, 186(14), pp. 4476-4485.

Camilli, A. and Bassler, B. L. (2006). Bacterial Small-Molecule Signalling Pathways. *Science*, 311(5764), pp. 1113-1116.

Castegnier, F., Ross, N., Chapuis, R., Deschênes, L. and Samson, R. (2006). Long-Term Persistence of a Nutrient-Starved Biofilm in a Limestone Fracture. *Water Research*, 40(5), pp.925-934.

Characklis, W. G., McFeters, G. A. and Marshall, K. C. (1990) Physiological Ecology in Biofilm Systems. *Biofilms*. New York, USA: John Wiley & Sons, pp. 341-394.

Choi, A. H., Slamti, L., Avci, F. Y., Pier, G. B. and Maira-Litrán, T. (2009). The *pgaABCD* Locus of *Acinetobacter baumannii* Encodes the Production of Poly-B-1-6-N-Acetylglucosamine, Which is Critical for Biofilm Formation. *Journal of Bacteriology*, 191(19), pp. 5953-5963.

Ciofu, O., Giwercman, B., Hoiby, N. and Pedersen, S. (2009). Development of Antibiotic Resistance in *Pseudomonas aeruginosa* during Two Decades of Antipseudomonal Treatment at the Danish CF Center. *APMIS* 102, pp. 674-680.

Ciofu, O., Tolker-Nielsen, T., Jensen, P., Wang, H. and Høiby, N. (2015). Antimicrobial Resistance, Respiratory Tract Infections and Role of Biofilms in Lung Infections in Cystic Fibrosis Patients. *Advanced Drug Delivery Reviews*, 85, pp. 7-23.

Costerton, J. W. (1999). Introduction to Biofilm. *International Journal of Antimicrobial Agents*, 11(3), pp. 217-221.

Costerton, J. W., Lewandowski, Z., Caldwell, D. E., Korber, D. R. and Lappin-Scott, H. M. (1995). Microbial Biofilms. *Annual Reviews in Microbiology*, 49(1), pp. 711-745.

Costerton, J. W., Stewart, P. S. and Greenberg, E. P. (1999). Bacterial Biofilms: A Common Cause of Persistent Infections. *Science*, 284(5418), pp. 1318-1322.

Da Silva, M. E. Z., Camargo Filho, I., Endo, E. H., Nakamura, C. V., Ueda-Nakamura, T. and Dias Filho, B. P. (2008). Characterisation of Potential Virulence Markers in *Pseudomonas aeruginosa* Isolated from Drinking Water. *Antonie van Leeuwenhoek*, 93(4), pp. 323-334.

Dale, K., Wolfe, R., Sinclair, M., Hellard, M. and Leder, K. (2009). Sporadic Gastroenteritis and Recreational Swimming in a Longitudinal Community Cohort Study in Melbourne, Australia. *American Journal of Epidemiology*, 170(12), pp. 1469-1477.

Danese, P. N., Pratt, L. A. and Kolter, R. (2000). Exopolysaccharide Production is Required for Development of *Escherichia coli* K-12 Biofilm Architecture. *Journal of Bacteriology*, 182(12), pp. 3593-3596.

Davey, M. E. and O'Toole, G. A. (2000). Microbial Biofilms: From Ecology to Molecular Genetics. *Microbiology and Molecular Biology Reviews*, 64(4), pp. 847-867.

Davies, D. G., Parsek, M. R., Pearson, J. P., Iglewski, B. H., Costerton, J. W. and Greenberg, E. P. (1998). The Involvement of Cell-To-Cell

Signals in the Development of a Bacterial Biofilm. *Science*, 280(5361), pp. 295-298.

De Kievit, T. R., Parkins, M. D., Gillis, R. J., Srikumar, R., Ceri, H., Poole, K., Iglewski, B. H. and Storey, D. G. (2001). Multidrug Efflux Pumps: Expression Patterns and Contribution to Antibiotic Resistance in *Pseudomonas aeruginosa* Biofilms. *Antimicrobial Agents and Chemotherapy*, 45(6), pp. 1761-1770.

Deschaine, B., Heysel, A., Lenhart, B. and Murphy, H. (2018). Biofilm Formation and Toxin Production Provide a Fitness Advantage in Mixed Colonies of Environmental Yeast Isolates. *Ecology and Evolution*, 8(11), pp. 5541-5550.

Diggle, S. P., Stacey, R. E., Dodd, C., Cámara, M., Williams, P. and Winzer, K. (2006). The Galactophilic Lectin, *LecA*, Contributes to Biofilm Development in *Pseudomonas aeruginosa*. *Environmental Microbiology*, 8(6), pp. 1095-1104.

Donlan, R. M. (2001). Biofilm Formation: A Clinically Relevant Microbiological Process. *Clinical Infectious Diseases*, 33(8), pp. 1387-1392.

Donlan, R. M. and Costerton, J. W. (2002). Biofilms: Survival Mechanisms of Clinically Relevant Microorganisms. *Clinical Microbiology Reviews*, 15(2), pp. 167-193.

Drenkard, E. and Ausubel, F. M. (2002). *Pseudomonas* Biofilm Formation and Antibiotic Resistance are Linked to Phenotypic Variation. *Nature*, 416(6882), pp. 740-743.

Elkins, J. G., Hassett, D. J., Stewart, P. S., Schweizer, H. P. and McDermott, T. R. (1999). Protective Role of Catalase in *Pseudomonas aeruginosa* Biofilm Resistance to Hydrogen Peroxide. *Applied and Environmental Microbiology*, 65(10), pp. 4594-4600.

Engel, L. S., Hill, J. M., Moreau, J. M., Green, L. C., Hobden, J. A. and O'Callaghan, R. J. (1998). *Pseudomonas aeruginosa* Protease IV Produces Corneal Damage and Contributes to Bacterial Virulence. *Investigative Ophthalmology and Visual Science*, 39(3), pp. 662-665.

Evans, L. R. and Linker, A., 1973. Production and Characterization of the Slime Polysaccharide of *Pseudomonas aeruginosa*. *Journal of Bacteriology*, 116(2), pp. 915-924.

Filho, N. A., Aires, M., Braz, D., Hinrichs, R., Macedo, A. and Alves Jr, C. (2018). Titanium Surface Chemical Composition Interferes in the *Pseudomonas aeruginosa* Biofilm Formation. *Artificial Organs*, 42(2), pp. 193-199.

Fito-Boncompte, L., Chapalain, A., Bouffartigues, E., Chaker, H., Lesouhaitier, O., Gicquel, G., Bazire, A., Madi, A., Connil, N., Véron, W. and Taupin, L. (2011). Full Virulence of *Pseudomonas aeruginosa* Requires *OprF*. *Infection and Immunity*, 79(3), pp. 1176-1186.

Fletcher, M. (1996). Bacterial Attachment in Aquatic Environments: A Diversity of Surfaces and Adhesion Strategies. In: M. Fletcher, ed., *Bacterial Adhesion: Molecular and Ecological Diversity*. New York: Wiley, pp. 1-24.

Fletcher, M. and Marshall, K. C. (1982). Bubble Contact Angle Method for Evaluating Substratum Interfacial Characteristics and Its Relevance to Bacterial Attachment. *Applied and Environmental Microbiology*, 44(1), pp. 184-192.

Friedman, L. and Kolter, R. (2003). Genes Involved in Matrix Formation in *Pseudomonas aeruginosa* PA14 Biofilms. *Molecular Microbiology*, 51(3), pp. 675-690.

Friedman, L. and Kolter, R. (2004). Two Genetic Loci Produce Distinct Carbohydrate-Rich Structural Components of the *Pseudomonas aeruginosa* Biofilm Matrix. *Journal of Bacteriology*, 186(14), pp. 4457-4465.

Fux, C. A., Costerton, J. W., Stewart, P. S. and Stoodley, P. (2005). Survival Strategies of Infectious Biofilms. *Trends in Microbiology*, 13(1), pp. 34-40.

Gabr, R. M., Gad-Elrab, S. M., Abskharon, R. N., Hassan, S. H. and Shoreit, A. A. (2009). Biosorption Of Hexavalent Chromium Using Biofilm of *E. coli* Supported on Granulated Activated Carbon. *World Journal of Microbiology and Biotechnology*, 25(10), pp. 1695-1703.

Geipel, U. (2009). Pathogenic Organisms in Hip Joint Infections. *International Journal of Medical Sciences*, 6(5), pp. 234-240.

Gellatly, S. L. and Hancock, R. E. (2013). *Pseudomonas aeruginosa*: New Insights into Pathogenesis and Host Defenses. *Pathogens and Disease*, 67(3), pp. 159-173.

Gening, M. L., Maira-Litrán, T., Kropec, A., Skurnik, D., Grout, M., Tsvetkov, Y. E., Nifantiev, N. E. and Pier, G. B. (2010). Synthetic B-(1→6)-Linked N-Acetylated and Nonacetylated Oligoglucosamines Used to Produce Conjugate Vaccines for Bacterial Pathogens. *Infection and Immunity*, 78(2), pp. 764-772.

Gerstel, U., Park, C. and Römling, U. (2004). Complex Regulation of *csgD* Promoter Activity by Global Regulatory Proteins. *Molecular Microbiology*, 49(3), pp. 639-654.

Gilbert, P., Das, J. and Foley, I. (1997). Biofilm Susceptibility to Antimicrobials. *Advances in Dental Research*, 11(1), pp. 160-167.

Giwercman, B., Jensen, E. T., Høiby, N., Kharazmi, A. and Costerton, J. W. (1991). Induction of Beta-Lactamase Production in *Pseudomonas aeruginosa* Biofilm. *Antimicrobial Agents and Chemotherapy*, 35(5), pp. 1008-1010.

Goosney, D. L., DeVinney, R., Pfuetzner, R. A., Frey, E. A., Strynadka, N. C. and Finlay, B. B. (2000). Enteropathogenic *E. coli* Translocated Intimin Receptor, *Tir*, Interacts Directly With α-Actinin. *Current Biology*, 10(12), pp. 735-738.

Gottenbos, B. (2001). Antimicrobial Effects of Positively Charged Surfaces on Adhering Gram-positive and Gram-negative Bacteria. *Journal of Antimicrobial Chemotherapy*, 48(1), pp.7-13.

Götz, F. (2002). *Staphylococcus* and Biofilms. *Molecular Microbiology*, 43(6), pp. 1367-1378.

Govan, J. R. and Deretic, V. (1996). Microbial Pathogenesis in Cystic Fibrosis: Mucoid *Pseudomonas aeruginosa* and *Burkholderia cepacia*. *Microbiological Reviews*, 60(3), pp. 539-574.

Hall-Stoodley, L. and Stoodley, P. (2002). Developmental Regulation of Microbial Biofilms. *Current Opinion in Biotechnology*, 13(3), pp. 228-233.

Hall-Stoodley, L., Costerton, J. W. and Stoodley, P. (2004). Bacterial Biofilms: From the Natural Environment to Infectious Diseases. *Nature Reviews Microbiology*, 2(2), pp. 95-108.

Hall-Stoodley, L. and Stoodley, P. (2005). Biofilm Formation and Dispersal and the Transmission of Human Pathogens. *Trends in Microbiology*, 13(1), pp. 7-10.

Hammer, B. K. and Bassler, B. L. (2003). Quorum Sensing Controls Biofilm Formation in *Vibrio cholerae*. *Molecular Microbiology*, 50(1), pp.101-104.

Hancock, R. E. and Speert, D. P. (2000). Antibiotic Resistance in *Pseudomonas aeruginosa*: Mechanisms and Impact on Treatment. *Drug Resistance Updates*, 3(4), pp. 247-255.

Hauser, A. R. (2009). The Type III Secretion System of *Pseudomonas aeruginosa*: Infection by Injection. *Nature Reviews Microbiology*, 7(9), pp.654.

Hayes, E., Pohl, K., McElvaney, N. G. and Reeves, E. P. (2011). The Cystic Fibrosis Neutrophil: A Specialized yet Potentially Defective Cell. *Archivum Immunologiae et Therapiae Experimentalis*, 59(2), pp. 97-112.

Hawser, S. P. and Douglas, L. J. (1994). Biofilm Formation by *Candida* Species on the Surface of Catheter Materials In Vitro. *Infection and Immunity*, 62(3), pp. 915-921.

Hentzer, M., Teitzel, G. M., Balzer, G. J., Heydorn, A., Molin, S., Givskov, M. and Parsek, M. R. (2001). Alginate Overproduction Affects *Pseudomonas aeruginosa* Biofilm Structure and Function. *Journal of Bacteriology*, 183(18), pp. 5395-5401.

Highsmith, A. K., Le, P. N., Khabbaz, R. F. and Munn, V. P. (1985). Characteristics of *Pseudomonas aeruginosa* Isolated from Whirlpools and Bathers. *Infection Control and Hospital Epidemiology*, 6(10), pp. 407-412.

Hincapié, O., Giraldo, P. and Orduz, S. (2018). In Silico Design of Polycationic Antimicrobial Peptides Active Against *Pseudomonas aeruginosa* and *Staphylococcus aureus*. *Antonie van Leeuwenhoek*, 111(10), pp. 1871-1882.

Høiby, N., Bjarnsholt, T., Givskov, M., Molin, S. and Ciofu, O. (2010). Antibiotic Resistance of Bacterial Biofilms. *International Journal of Antimicrobial Agents*, 35(4), pp. 322-332.

Horke, S., Witte, I., Altenhöfer, S., Wilgenbus, P., Goldeck, M., Förstermann, U., Xiao, J., Kramer, G. L., Haines, D. C., Chowdhary, P. K. and Haley, R. W. (2010). Paraoxonase 2 is Down-regulated by the *Pseudomonas aeruginosa* Quorum Sensing Signal N-(3-Oxododecanoyl)-L-Homoserine Lactone and Attenuates Oxidative Stress Induced by Pyocyanin. *Biochemical Journal*, 426(1), pp. 73-83.

Isberg, R. R. and Barnes, P. (2001). Subversion of Integrins by Enteropathogenic *Yersinia*. *Journal of Cell Science*, 114(1), pp. 21-28.

Itoh, Y., Rice, J. D., Goller, C., Pannuri, A., Taylor, J., Meisner, J., Beveridge, T. J., Preston, J. F. and Romeo, T. (2008). Roles of *PgaABCD* Genes in Synthesis, Modification, and Export of the *Escherichia coli* Biofilm Adhesin Poly-B-1, 6-N-Acetyl-D-Glucosamine. *Journal of Bacteriology*, 190(10), pp. 3670-3680.

Itoh, Y., Wang, X., Hinnebusch, B. J., Preston III, J. F., Romeo, T. (2005) Depolymerization of β-1, 6-N-Acetyl-d-Glucosamine Disrupts the Integrity of Diverse Bacterial Biofilms. *Journal of Bacteriology* 187: 382-387.

Izano, E. A., Amarante, M. A., Kher, W. B. and Kaplan, J. B. (2008). Differential Roles of Poly-N-Acetylglucosamine Surface Polysaccharide and Extracellular DNA in *Staphylococcus aureus* and *Staphylococcus epidermidis* Biofilms. *Applied and Environmental Microbiology*, 74(2), pp. 470-476.

Isberg, R. R. and van Nhieu, G. T. (1995). The Mechanism of Phagocytic Uptake Promoted By Invasin-Integrin Interaction. *Trends in Cell Biology*, 5(3), pp. 120-124.

Isberg, R. R., Voorhis, D. L. and Falkow, S. (1987). Identification of Invasin: A Protein that Allows Enteric Bacteria to Penetrate Cultured Mammalian Cells. *Cell*, 50(5), pp. 769-778.

Ishida, H., Ishida, Y., Kurosaka, Y., Otani, T., Sato, K. and Kobayashi, H. (1998). In Vitro and In Vivo Activities Of Levofloxacin Against

Biofilm-Producing *Pseudomonas aeruginosa*. *Antimicrobial Agents and Chemotherapy*, 42(7), pp. 1641-1645.

Jacoby, G. A. (2009). *AmpC* β-lactamases. *Clinical Microbiology Reviews*, 22(1), pp. 161-182.

Jackson, D. W., Suzuki, K., Oakford, L., Simecka, J. W., Hart, M. E. and Romeo, T. (2002). Biofilm Formation and Dispersal Under the Influence of the Global Regulator *CsrA* of *Escherichia coli*. *Journal of Bacteriology*, 184(1), pp. 290-301.

Jolivet-Gougeon, A. and Bonnaure-Mallet, M. (2014). Biofilms as a Mechanism of Bacterial Resistance. *Drug Discovery Today: Technologies*, 11, pp. 49-56.

Jones, S. M., Yerly, J., Hu, Y., Ceri, H. and Martinuzzi, R. (2007). Structure of *Proteus mirabilis* Biofilms Grown in Artificial Urine and Standard Laboratory Media. *FEMS Microbiology Letters*, 268(1), pp. 16-21.

Kadouri, D., & O'Toole, G. (2005). Susceptibility of Biofilms to *Bdellovibrio bacteriovorus* Attack. *Applied and Environmental Microbiology*, 71(7), 4044-4051.

Kang, Y., Liu, H., Genin, S., Schell, M. A. and Denny, T. P. (2002). *Ralstonia solanacearum* Requires Type 4 Pili to Adhere to Multiple Surfaces and for Natural Transformation and Virulence. *Molecular Microbiology*, 46(2), pp. 427-437.

Karatan, E. and Watnick, P. (2009). Signals, Regulatory Networks, and Materials that Build and Break Bacterial Biofilms. *Microbiology and Molecular Biology Reviews*, 73(2), pp. 310-347.

Katsikogianni, M. and Missirlis, Y. F. (2004). Concise Review of Mechanisms of Bacterial Adhesion to Biomaterials and of Techniques Used in Estimating Bacteria-Material Interactions. *European Cells Matererilas*, 8(3), pp. 37-57.

Kierek, K. and Watnick, P. I. (2003). The *Vibrio cholerae* O139 O-Antigen Polysaccharide is Essential for Ca2+-Dependent Biofilm Development in Sea Water. *Proceedings of the National Academy of Sciences*, 100(24), pp. 14357-14362.

King, J. D., Kocíncová, D., Westman, E. L. and Lam, J. S. (2009). Lipopolysaccharide Biosynthesis in *Pseudomonas aeruginosa*. *Innate Immunity*, 15(5), pp. 261-312.

Kipnis, E., Sawa, T. and Wiener-Kronish, J. (2006). Targeting Mechanisms of *Pseudomonas aeruginosa* Pathogenesis. *Medecine et Maladies Infectieuses*, 36(2), pp. 78-91.

Kirov, S. M., Castrisios, M. and Shaw, J. G. (2004). *Aeromonas* Flagella (Polar and Lateral) are Enterocyte Adhesins that Contribute to Biofilm Formation on Surfaces. *Infection and Immunity*, 72(4), pp. 1939-1945.

Kjelleberg, S. and Molin, S. (2002). Is There a Role for Quorum Sensing Signals in Bacterial Biofilms? *Current Opinion in Microbiology*, 5(3), pp. 254-258.

Klausen, M., Aaes-Jørgensen, A., Molin, S. and Tolker-Nielsen, T. (2003a). Involvement of Bacterial Migration in the Development of Complex Multicellular Structures in *Pseudomonas aeruginosa* Biofilms. *Molecular Microbiology*, 50(1), pp. 61-68.

Klausen, M., Heydorn, A., Ragas, P., Lambertsen, L., Aaes-Jørgensen, A., Molin, S. and Tolker-Nielsen, T. (2003b). Biofilm Formation by *Pseudomonas aeruginosa* Wild Type, Flagella and Type IV Pili Mutants. *Molecular Microbiology*, 48(6), pp. 1511-1524.

Koch, C. and Høiby, N. (1993). Pathogenesis of Cystic Fibrosis. *Lancet* 341: 1065.

Kuang, Z., Hao, Y., Walling, B. E., Jeffries, J. L., Ohman, D. E. and Lau, G. W. (2011). *Pseudomonas aeruginosa* Elastase Provides an Escape from Phagocytosis by Degrading the Pulmonary Surfactant Protein-A. *PloS One*, 6(11), pp. e27091.

Laarman, A. J., Bardoel, B. W., Ruyken, M., Fernie, J., Milder, F. J., van Strijp, J. A. and Rooijakkers, S. H. (2012). *Pseudomonas aeruginosa* Alkaline Protease Blocks Complement Activation via the Classical and Lectin Pathways. *The Journal of Immunology*, 188(1), pp. 386-393.

Lai, C. (2018). Bacterial Attachment, Aggregation, and Alignment on Subcellular Nanogratings. *Langmuir*, 34(13), pp. 4059-4070.

Lam, J., Chan, R., Lam, K. and Costerton, J. W. (1980). Production of Mucoid Microcolonies by *Pseudomonas aeruginosa* Within Infected Lungs in Cystic Fibrosis. *Infection and Immunity*, 28(2), pp. 546-556.

Lam, S., O'Brien-Simpson, N., Pantarat, N., Sulistio, A., Wong, E., Chen, Y., Lenzo, J., Holden, J., Blencowe, A., Reynolds, E. and Qiao, G. (2016). Combating Multidrug-resistant Gram-negative Bacteria with Structurally Nanoengineered Antimicrobial Peptide Polymers. *Nature Microbiology*, 1(11), pp. 1-11.

Lau, G. W., Hassett, D. J., Ran, H. and Kong, F. (2004). The Role of Pyocyanin in *Pseudomonas aeruginosa* Infection. *Trends in Molecular Medicine*, 10(12), pp. 599-606.

Lauriano, C. M., Ghosh, C., Correa, N. E. and Klose, K. E. (2004). The Sodium-Driven Flagellar Motor Controls Exopolysaccharide Expression in *Vibrio cholerae*. *Journal of Bacteriology*, 186(15), pp. 4864-4874.

Lemon, K. P., Higgins, D. E. and Kolter, R. (2007). Flagellar Motility is Critical for *Listeria monocytogenes* Biofilm Formation. *Journal of Bacteriology*, 189(12), pp. 4418-4424.

Leoni, E., Legnani, P., Guberti, E. and Masotti, A. (1999). Risk of Infection Associated with Microbiological Quality of Public Swimming Pools in Bologna, Italy. *Public Health*, 113(5), pp. 227-232.

Lewis, K. (2008). Multidrug tolerance of biofilms and persister cells. In: T. Romeo, ed., *Bacterial Biofilms*. Heidelberg: Springer, pp. 107-131.

Li, X. Z., Nikaido, H. and Poole, K. (1995). Role of *MexA-MexB-OprM* in Antibiotic Efflux in *Pseudomonas aeruginosa*. *Antimicrobial Agents and Chemotherapy*, 39(9), pp. 1948-1953.

Ling, G. C., Low, M. H., Erken, M., Longford, S., Nielsen, S., Poole, A. J., ... & Kjelleberg, S. (2014). Micro-fabricated Polydimethyl Siloxane (PDMS) Surfaces Regulate the Development of Marine Microbial Biofilm Communities. *Biofouling*, 30(3), 323-335.

Liu, Y. and Tay, J. H. (2002). The Essential Role of Hydrodynamic Shear Force in the Formation of Biofilm and Granular Sludge. *Water Research*, 36(7), pp. 1653-1665.

Livermore, D. M. (2002). Multiple Mechanisms of Antimicrobial Resistance in Pseudomonas aeruginosa: Our Worst Nightmare? *Clinical Infectious Diseases*, 34(5), pp. 634-640.

Luef, B., Neu, T. R., ZweimÜLler, I. and Peduzzi, P. (2009). Structure and Composition of Aggregates in Two Large European Rivers, Based on Confocal Laser Scanning Microscopy and Image and Statistical Analyses. *Applied and Environmental Microbiology*, 75(18), pp. 5952-5962.

Lutz, J. K. and Lee, J. (2011). Prevalence and Antimicrobial-Resistance of *Pseudomonas aeruginosa* in Swimming Pools and Hot Tubs. *International Journal of Environmental Research and Public Health*, 8(2), pp. 554-564.

Lyczak, J., Cannon, C. and Pier, G. (2002). Lung Infections Associated with Cystic Fibrosis. *Clinical Microbiology Reviews*, 15(2), pp. 194-222.

Ma, L., Jackson, K. D.,, Landry, R. M., Parsek, M. R. and Wozniak, D. J., (2006) Analysis of *Pseudomonas aeruginosa* conditional *Psl* variants reveals roles for the *Psl* polysaccharide in adhesion and maintaining biofilm structure post attachment. *Journal of Bacteriology* 188: 8213-8221.

Mah, T. F. C. and O'Toole, G. A. (2001). Mechanisms of Biofilm Resistance to Antimicrobial Agents. *Trends in Microbiology*, 9(1), pp. 34-39.

Marrie, T. J., Nelligan, J. and Costerton, J. W. (1982). A Scanning and Transmission Electron Microscopic Study of an Infected Endocardial Pacemaker Lead. *Circulation*, 66(6), pp. 1339-1341.

Massol-Deyá, A. A., Whallon, J., Hickey, R. F. and Tiedje, J. M. (1995). Channel Structures in Aerobic Biofilms of Fixed-Film Reactors Treating Contaminated Groundwater. *Applied and Environmental Microbiology*, 61(2), pp. 769-777.

Mathee, K., Ciofu, O., Sternberg, C., Lindum, P. W., Campbell, J. I., Jensen, P., Johnsen, A. H., Givskov, M., Ohman, D. E., Søren, M. and Høiby, N. (1999). Mucoid Conversion of *Pseudomonas aeruginosa* by

Hydrogen Peroxide: A Mechanism for Virulence Activation in the Cystic Fibrosis Lung. *Microbiology*, 145(6), pp. 1349-1357.

Matsukawa, M. and Greenberg, E. P. (2004). Putative Exopolysaccharide Synthesis Genes Influence *Pseudomonas aeruginosa* Biofilm Development. *Journal of Bacteriology*, 186(14), pp. 4449-4456.

Mattick, J. S. (2002). Type IV Pili and Twitching Motility. *Annual Reviews in Microbiology*, 56(1), pp. 289-314.

Matz, C., McDougald, D., Moreno, A., Yung, P., Yildiz, F. and Kjelleberg, S. (2005). Biofilm Formation and Phenotypic Variation Enhance Predation-Driven Persistence of *Vibrio cholerae. Proceedings of the National Academy of Sciences*, 102(46), pp. 16819-16824.

Matz, C., Webb, J., Schupp, P., Phang, S., Penesyan, A., Egan, S., Steinberg, P. and Kjelleberg, S. (2008). Marine Biofilm Bacteria Evade Eukaryotic Predation by Targeted Chemical Defense. *PLoS ONE*, 3(7), pp. e2744.

Mcclaine, J. W. and Ford, R. M. (2002). Characterizing the Adhesion of Motile and Nonmotile *Escherichia coli* to a Glass Surface Using a Parallel-Plate Flow Chamber. *Biotechnology and Bioengineering*, 78(2), pp. 179-189.

Mcdougald, D., Rice, S. A., Barraud, N., Steinberg, P. D. and Kjelleberg, S. (2011). Should We Stay or Should We Go: Mechanisms and Ecological Consequences for Biofilm Dispersal. *Nature Reviews Microbiology*, 10(1), pp. 39-50.

Mehanna, M., BassÉGuy, R., DÉLia, M. L. and Bergel, A. (2009). Effect of *Geobacter sulfurreducens* on the Microbial Corrosion of Mild Steel, Ferritic and Austenitic Stainless Steels. *Corrosion Science,* 51(11), pp. 2596-2604.

Meibom, K. L., Li, X. B., Nielsen, A. T., Wu, C. Y., Roseman, S. and Schoolnik, G. K. (2004). The *Vibrio cholerae* Chitin Utilization Program. *Proceedings of the National Academy of Sciences*, 101(8), pp. 2524-2529.

Mena, K. D. and Gerba, C. P. (2009). Risk assessment of *Pseudomonas aeruginosa* in water. In: D. Whitacre, ed., *Reviews of Environmental Contamination and Toxicology* Vol 201. Boston: Springer, pp. 71-115.

Mengaud, J., Ohayon, H., Gounon, P., MÈGe, R. M. and Cossart, P. (1996). E-Cadherin is the Receptor for Internalin, a Surface Protein Required for Entry of *L. monocytogenes* into Epithelial Cells. *Cell*, 84(6), pp. 923-932.

Merritt, P. M., Danhorn, T. and Fuqua, C. (2007). Motility and Chemotaxis in *Agrobacterium tumefaciens* Surface Attachment and Biofilm Formation. *Journal of Bacteriology*, 189(22), pp. 8005-8014.

Miao, E. A., Andersen-Nissen, E., Warren, S. E. and Aderem, A. (2007). *TLR5* and *Ipaf*: Dual Sensors of Bacterial Flagellin in the Innate Immune System. *Seminars in Immunopathology*, 29(3), pp. 275-288.

Mitik-Dineva, N., Wang, J., Truong, V., Stoddart, P., Malherbe, F., Crawford, R. and Ivanova, E. (2008). *Escherichia coli*, *Pseudomonas aeruginosa*, and *Staphylococcus aureus* Attachment Patterns on Glass Surfaces with Nanoscale Roughness. *Current Microbiology*, 58(3), pp. 268-273.

Mittal, R., Aggarwal, S., Sharma, S., Chhibber, S. and Harjai, K. (2009). Urinary Tract Infections Caused by *Pseudomonas aeruginosa*: A Minireview. *Journal of Infection and Public Health*, 2(3), pp. 101-111.

Molin, S. and Tolker-Nielsen, T. (2003). Gene Transfer Occurs with Enhanced Efficiency in Biofilms and Induces Enhanced Stabilisation of the Biofilm Structure. *Current Opinion in Biotechnology*, 14(3), pp. 255-261.

Mulligan, M. E., Murray-Leisure, K. A., Ribner, B. S., Standiford, H. C., John, J. F., Korvick, J. A., Kauffman, C. A. and Victor, L. Y. (1993). Methicillin-Resistant *Staphylococcus aureus*: A Consensus Review of the Microbiology, Pathogenesis, and Epidemiology with Implications for Prevention and Management. *The American Journal of Medicine*, 94(3), pp. 313-328.

Nadell, C., Drescher, K. and Foster, K. (2016). Spatial Structure, Cooperation and Competition in Biofilms. *Nature Reviews Microbiology*, 14(9), pp. 589-600.

Nakajima, A., Sugimoto, Y., Yoneyama, H. and Nakae, T. (2002). High-Level Fluoroquinolone Resistance in *Pseudomonas aeruginosa*

Due to Interplay of the *MexAB-OprM* Efflux Pump and the DNA Gyrase Mutation. *Microbiology and Immunology*, 46(6), pp. 391-395.

Nemoto, K., Hirota, K., Murakami, K., Taniguti, K., Murata, H., Viducic, D. and Miyake, Y. (2003). Effect of Varidase (Streptodornase) on Biofilm Formed by *Pseudomonas aeruginosa*. *Chemotherapy*, 49(3), pp. 121-125.

Neu, T. R., & Lawrence, J. R. (1997). Development and Structure of Microbial Biofilms in River Water Studied by Confocal Laser Scanning Microscopy. *FEMS Microbiology Ecology*, 24(1), 11-25.

Nield-Gehrig, J. S. (2005). Dental Plaque Biofilms. *Journal of Practical Hygiene*, 14(16), pp. 1-6.

O'Toole, G. A. and Kolter, R. (1998). Flagellar and Twitching Motility are Necessary for *Pseudomonas aeruginosa* Biofilm Development. *Molecular Microbiology*, 30(2), pp. 295-304.

O'Toole, G. A., Kaplan, H. B. and Kolter, R. (2000). Biofilm Formation as Microbial Development. *Annual Reviews in Microbiology*, 54(1), pp. 49-79.

O'Toole, G. A. and Kolter, R. (2002). Initiation of Biofilm Formation in *Pseudomonas fluorescens* WCS365 Proceeds via Multiple, Convergent Signalling Pathways: A Genetic Analysis. *Molecular Microbiology* 28, pp. 449-461.

Pamp, S. J., Gjermansen, M., Johansen, H. K. and Tolker-Nielsen, T. (2008). Tolerance to the Antimicrobial Peptide Colistin in *Pseudomonas aeruginosa* Biofilms is Linked to Metabolically Active Cells, and Depends on the *pmr* and *mexAB-oprM* Genes. *Molecular Microbiology*, 68(1), pp. 223-240.

Pendleton, J., Gorman, S. and Gilmore, B. (2013). Clinical Relevance of the ESKAPE Pathogens. *Expert Review of Anti-infective Therapy*, 11(3), pp. 297-308.

Peters, G., Locci, R. and Pulverer, G. (1981). Microbial Colonization of Prosthetic Devices. II. Scanning Electron Microscopy of Naturally Infected Intravenous Catheters. *Zentralblatt fur Bakteriologie, Mikrobiologie und Hygiene. 1. Abt. Originale B, Hygiene*, 173(5), pp. 293-299.

Peters, G., Locci, R. and Pulverer, G. (1982). Adherence and Growth of Coagulase-Negative *Staphylococci* on Surfaces of Intravenous Catheters. *Journal of Infectious Diseases*, 146(4), pp. 479-482.

Powers, R. A., BlÁZquez, J., Weston, G. S., Morosini, M. I., Baquero, F. and Shoichet, B. K. (1999). The Complexed Structure and Antimicrobial Activity of a Non-β-Lactam Inhibitor of AmpC β-Lactamase. *Protein Science*, 8(11), pp. 2330-2337.

Pratt, L. A. and Kolter, R. (2002). Genetic Analysis of *Escherichia coli* Biofilm Formation: Roles of Flagella, Motility, Chemotaxis and Type I Pili. *Molecular Microbiology*, 30(2), pp. 285-293.

Price, D. and Ahearn, D. G. (1988). Incidence and Persistence of *Pseudomonas aeruginosa* In Whirlpools. *Journal of Clinical Microbiology*, 26(9), pp. 1650-1654.

Prigent-Combaret, C., Prensier, G., Le Thi, T. T., Vidal, O., Lejeune, P. and Dorel, C. (2001). Developmental Pathway for Biofilm Formation in Curli-Producing *Escherichia coli* Strains: Role of Flagella, Curli and Colanic Acid. *Environmental Microbiology*, 2(4), pp. 450-464.

Purevdorj, B., Costerton, J. W. and Stoodley, P. (2002). Influence of Hydrodynamics and Cell Signalling on the Structure and Behavior of *Pseudomonas aeruginosa* Biofilms. *Applied and Environmental Microbiology*, 68(9), pp. 4457-4464.

Quirynen, M. and Bollen, C. M. L. (1995). The Influence of Surface Roughness and Surface-Free Energy on Supra-and Subgingival Plaque Formation in Man: A Review of the Literature. *Journal of Clinical Periodontology*, 22(1), pp. 1-14.

Rabaey, K., Rodriguez, J., Blackall, L. L., Keller, J., Gross, P., Batstone, D., Verstraete, W. and Nealson, K. H. (2007). Microbial Ecology Meets Electrochemistry: Electricity-Driven and Driving Communities. *The ISME Journal*, 1(1), pp. 9-18.

Rice, S. A., Van den Akker, B., Pomati, F. and Roser, D. (2012). A Risk Assessment of *Pseudomonas aeruginosa* In Swimming Pools: A Review. *Journal of Water and Health*, 10(2), pp. 181-196.

Roger, P., Puchelle, E., Bajolet-Laudinat, O., Tournier, J. M., Debordeaux, C., Plotkowski, M. C., Cohen, J. M., Sheppard, D. and De Bentzmann,

S. (1999). Fibronectin and Alpha5beta1 Integrin Mediate Binding of *Pseudomonas aeruginosa* to Repairing Airway Epithelium. *European Respiratory Journal*, 13(6), pp. 1301-1309.

Saiman, L. and Prince, A. (1993). *Pseudomonas aeruginosa* Pili Bind to asialoGM1 Which is Increased on the Surface of Cystic Fibrosis Epithelial Cells. *The Journal of Clinical Investigation*, 92(4), pp. 1875-1880.

Salmen, P., Dwyer, M. D., Vorse, M. D. and Kruse, M. D. (1983). Whirlpool-Associated *Pseudomonas aeruginosa*. *Journal of American Medical Association*, 250, pp. 2025-2026.

Sanchez, C., Mende, K., Beckius, M., Akers, K., Romano, D., Wenke, J. and Murray, C. (2013). Biofilm Formation by Clinical Isolates and the Implications in Chronic Infections. *BMC Infectious Diseases*, 13(1), pp. 47-59.

Sandle, T. (2013). Bacterial Adhesion: An Introduction. *Journal of Validation Technology*, 19(9), pp. 1-10.

Schleheck, D., Barraud, N., Klebensberger, J., Webb, J. S., Mcdougald, D., Rice, S. A. and Kjelleberg, S. (2009). *Pseudomonas aeruginosa* PAO1 Preferentially Grows as Aggregates in Liquid Batch Cultures and Disperses upon Starvation. *PloS One*, 4(5), pp. E5513.

Shigeta, M., Tanaka, G., Komatsuzawa, H., Sugai, M., Suginaka, H. and Usui, T. (1997). Permeation of Antimicrobial Agents through *Pseudomonas aeruginosa* Biofilms: A Simple Method. *Chemotherapy*, 43(5), pp. 340-345.

Shikuma, N. J. and Hadfield, M. G. (2010). Marine Biofilms on Submerged Surfaces are a Reservoir for *Escherichia coli* and *Vibrio cholerae*. *Biofouling*, 26(1), pp. 39-46.

Singh, R., Paul, D. and Jain, R. K. (2006). Biofilms: Implications in Bioremediation. *Trends in Microbiology*, 14(9), pp. 389-397.

Singh, P. K., Schaefer, A. L., Parsek, M. R., Moninger, T. O., Welsh, M. J. and Greenberg, E. P. (2000). Quorum-Sensing Signals Indicate that Cystic Fibrosis Lungs are Infected with Bacterial Biofilms. *Nature*, 407(6805), pp. 762-764.

Sinha, B., François, P. P., Nüße, O., Foti, M., Hartford, O. M., Vaudaux, P., Foster, T. J., Lew, D. P., Herrmann, M. and Krause, K. H. (2001). Fibronectin-Binding Protein Acts as *Staphylococcus aureus* Invasin via Fibronectin Bridging to Integrin A5B1. *Cellular Microbiology*, 1(2), pp. 101-117.

Smith, R. S., Harris, S. G., Phipps, R. and Iglewski, B. (2002). The *Pseudomonas aeruginosa* Quorum-Sensing Molecule N-(3-Oxododecanoyl) Homoserine Lactone Contributes to Virulence and Induces Inflammation In Vivo. *Journal of Bacteriology*, 184(4), pp. 1132-1139.

Schneider-Futschik, E., Paulin, O., Hoyer, D., Roberts, K., Ziogas, J., Baker, M., Karas, J., Li, J. and Velkov, T. (2018). Sputum Active Polymyxin Lipopeptides: Activity against Cystic Fibrosis *Pseudomonas aeruginosa* Isolates and Their Interactions with Sputum Biomolecules. *ACS Infectious Diseases*, 4(5), pp. 646-655.

Stanley, N. R. and Lazazzera, B. A. (2004). Environmental Signals and Regulatory Pathways that Influence Biofilm Formation. *Molecular Microbiology*, 52(4), pp. 917-924.

Stoodley, P., Sauer, K., Davies, D. G. and Costerton, J. W. (2002). Biofilms as Complex Differentiated Communities. *Annual Reviews in Microbiology*, 56(1), pp. 187-209.

Sun, S., Noorian, P. and McDougald, D. (2018). Dual Role of Mechanisms Involved in Resistance to Predation by Protozoa and Virulence to Humans. *Frontiers in Microbiology*, 9, pp. 1017.

Suter, S., Schaad, U. B., Morgenthaler, J. J., Chevallier, I. and Schnebli, H. P. (1988). Fibronectin-Cleaving Activity in Bronchial Secretions of Patients with Cystic Fibrosis. *Journal of Infectious Diseases*, 158(1), pp. 89-100.

Sutherland, I. W. (2001a). The Biofilm Matrix–an Immobilized but Dynamic Microbial Environment. *Trends in Microbiology*, 9(5), pp. 222-227.

Sutherland, I. W. (2001b). Biofilm Exopolysaccharides: A Strong and Sticky Framework. *Microbiology*, 147(1), pp. 3-9.

Tate, D., Mawer, S. and Newton, A. (2003). Outbreak of *Pseudomonas aeruginosa* Folliculitis Associated with a Swimming Pool Inflatable. *Epidemiology and Infection*, 130(2), pp. 187-192.

Taylor, P. K., Yeung, A. T. and Hancock, R. E. (2014). Antibiotic Resistance in *Pseudomonas aeruginosa* Biofilms: Towards the Development of Novel Anti-Biofilm Therapies. *Journal of Biotechnology*, 191, pp. 121-130.

Teghanemt, A., Zhang, D., Levis, E. N., Weiss, J. P. and Gioannini, T. L. (2005). Molecular Basis of Reduced Potency of Underacylated Endotoxins. *The Journal of Immunology*, 175(7), pp. 4669-4676.

Tielker, D., Hacker, S., Loris, R., Strathmann, M., Wingender, J., Wilhelm, S., Rosenau, F., and Jaeger, K. E. (2005). *Pseudomonas aeruginosa* Lectin *Lecb* from Located in the Outer Membrane and is Involved in Biofilm Formation. *Microbiology*, 151(5), pp. 1313-1323.

Thormann, K. M., Saville, R. M., Shukla, S., and Spormann, A. M. (2005). Induction of Rapid Detachment in *Shewanella oneidensis* MR-1 Biofilms. *Journal of Bacteriology*, 187(3), pp. 1014-1021.

Tolker-Nielsen, T., Brinch, U. C., Ragas, P. C., Andersen, J. B., Jacobsen, C. S., and Molin, S. (2000). Development and Dynamics of *Pseudomonas sp.* Biofilms. *Journal of Bacteriology*, 182(22), pp. 6482-6489.

Toutain, C. M., Caizza, N. C., Zegans, M. E., and O'Toole, G. A. (2007). Roles for Flagellar Stators in Biofilm Formation by *Pseudomonas aeruginosa*. *Research in Microbiology*, 158(5), pp. 471-477.

Uhl, W. and Hartmann, C. (2005). Disinfection By-Products and Microbial Contamination in the Treatment of Pool Water with Granular Activated Carbon. *Water Science and Technology*, 52(8), pp. 71-76.

Van Loosdrecht, M. C. M., and Zehnder, A. J. B. (1990). Energetics of Bacterial Adhesion. *Experientia*, 46(8), pp. 817-822.

Van Loosdrecht, M. C., Lyklema, J., Norde, W., and Zehnder, A. J. (1990a). Influence of Interfaces on Microbial Activity. *Microbiological Reviews*, 54(1), pp. 75-87.

Van Loosdrecht, M. C., Norde, W., Lyklema, J., and Zehnder, A. J. (1990b). Hydrophobic and Electrostatic Parameters in Bacterial Adhesion. *Aquatic Sciences*, 52(1), pp. 103-114.

Van Loosdrecht, M. C., Lyklema, J., Norde, W., Schraa, G., and Zehnder, A. J. (1987). Electrophoretic Mobility and Hydrophobicity as a Measured to Predict the Initial Steps of Bacterial Adhesion. *Applied and Environmental Microbiology*, 53(8), pp. 1898-1901.

Van Oss, C. J., Good, R. J., and Chaudhury, M. K. (1986). The Role of Van Der Waals Forces and Hydrogen Bonds in "Hydrophobic Interactions" between Biopolymers and Low Energy Surfaces. *Journal of Colloid and Interface Science*, 111(2), pp. 378-390.

Van Schaik, E. J., Giltner, C. L., Audette, G. F., Keizer, D. W., Bautista, D. L., Slupsky, C. M., Sykes, B. D., and Irvin, R. T. (2005). DNA Binding: A Novel Function of *Pseudomonas aeruginosa* Type IV Pili. *Journal of Bacteriology*, 187(4), pp. 1455-1464.

Vogne, C., Aires, J. R., Bailly, C., Hocquet, D. and Plésiat, P. (2004). Role of the Multidrug Efflux System *mexXY* in the Emergence of Moderate Resistance to Aminoglycosides among *Pseudomonas aeruginosa* Isolates from Patients with Cystic Fibrosis. *Antimicrobial Agents and Chemotherapy*, 48(5), pp. 1676-1680.

von Canstein, H., Ogawa, J., Shimizu, S., and Lloyd, J. R. (2008). Secretion of Flavins by *Shewanella species* and their Role in Extracellular Electron Transfer. *Applied and Environmental Microbiology*, 74(3), pp. 615-623.

Von Eiff, C., Heilmann, C., Herrmann, M. and Peters, G. (1999). Basic Aspects of the Pathogenesis of *Staphylococcal* Polymer-Associated Infections. *Infection*, 27, pp. S7-S10.

Wang, X., Preston, J. F., and Romeo, T. (2004). The *pgaabcd* Locus of *Escherichia coli* Promotes the Synthesis of a Polysaccharide Adhesin Required for Biofilm Formation. *Journal of Bacteriology*, 186(9), pp. 2724-2734.

Weitere, M., Bergfeld, T., Rice, S. A., Matz, C. and Kjelleberg, S. (2005). Grazing Resistance of *Pseudomonas aeruginosa* Biofilms Depends on

Type of Protective Mechanism, Developmental Stage and Protozoan Feeding Mode. *Environmental Microbiology*, 7(10), pp. 1593-1601.

Westall, F., De Wit, M. J., Dann, J., Van Der Gaast, S., De Ronde, C. E., and Gerneke, D. (2001). Early Archean Fossil Bacteria and Biofilms in Hydrothermally-Influenced Sediments from the Barberton Greenstone Belt, South Africa. *Precambrian Research*, 106(1-2), pp. 93-116.

Whitchurch, C. B., Tolker-Nielsen, T., Ragas, P. C., and Mattick, J. S. (2002). Extracellular DNA Required for Bacterial Biofilm Formation. *Science*, 295(5559), pp. 1487-1487.

Willcox, M. D. (2007). *Pseudomonas aeruginosa* Infection and Inflammation during Contact Lens Wear: A Review. *Optometry and Vision Science*, 84(4), pp. 273-278.

Wu, G., Nielsen, M., Sorensen, K., Zhan, X., and Rodgers, M. (2009). Distributions and Activities of Ammonia Oxidizing Bacteria and Polyphosphate Accumulating Organisms in a Pumped-Flow Biofilm Reactor. *Water Research*, 43(18), pp. 4599-4609.

Xu, K. D., Stewart, P. S., Xia, F., Huang, C. T. and McFeters, G. A. (1998). Spatial Physiological Heterogeneity in *Pseudomonas aeruginosa* Biofilm is Determined by Oxygen Availability. *Applied and Environmental Microbiology*, 64(10), pp. 4035-4039.

Yates, S. P., Taylor, P. L., Jørgensen, R., Ferraris, D., Zhang, J., Andersen, G. R. and Merrill, A. R. (2005). Structure–Function Analysis of Water-Soluble Inhibitors of the Catalytic Domain of Exotoxin A from *Pseudomonas aeruginosa*. *Biochemical Journal*, 385(3), pp. 667-675.

Yoon, S. S., Hennigan, R. F., Hilliard, G. M., Ochsner, U. A., Parvatiyar, K., Kamani, M. C., Allen, H. L., DeKievit, T. R., Gardner, P. R., Schwab, U. and Rowe, J. J. (2002). *Pseudomonas aeruginosa* Anaerobic Respiration in Biofilms: Relationships to Cystic Fibrosis Pathogenesis. *Developmental Cell*, 3(4), pp. 593-603.

Young, T. (1805). III. An Essay on the Cohesion of Fluids. *Philosophical Transactions of the Royal Society of London*, 95, pp. 65-87.

Yu, Y., Cheng, A. S., Wang, L., Dunne, W. M. and Bayliss, S. J. (2007). Hot Tub Folliculitis or Hot Hand–Foot Syndrome Caused by

Pseudomonas aeruginosa. *Journal of the American Academy of Dermatology*, 57(4), pp. 596-600.

Zaborina, O., Holbrook, C., Chen, Y., Long, J., Zaborin, A., Morozova, I., Fernandez, H., Wang, Y., Turner, J. and Alverdy, J. (2008). Structure–Function Aspects of *PstS* in Multi-Drug–Resistant *Pseudomonas aeruginosa*. *PLoS Pathogens*, 4(2), p. e43.

Zhang, W., Culley, D. E., Nie, L., and Scholten, J. C. (2007). Comparative Transcriptome Analysis of *Desulfovibrio vulgaris* Grown in Planktonic Culture and Mature Biofilm on a Steel Surface. *Applied Microbiology and Biotechnology*, 76(2), pp. 447-457.

In: *Pseudomonas aeruginosa* ISBN: 978-1-53616-100-7
Editors: C. Fah Wong and H. Idris

Chapter 3

QUORUM SENSING IN *Pseudomonas aeruginosa*: UNDERSTANDING AND INTERFERING POTENTIAL

***Hoai T. T. Nguyen*[1], *Ly T. T. Trinh*[1] *and Peter A. Lambert*[2]**
[1]School of Biotechnology, International University,
Vietnam National University HCMC, Ho Chi Minh, Vietnam
[2]School of Life and Health Sciences, Aston University,
Birmingham, UK

ABSTRACT

Quorum sensing (QS), an intercellular communication system, plays a key role in the regulation of genes encoding for biofilm formation and other extracellular virulence factors. This system allows bacteria to control gene expression in response not only to cell population density as originally described, but also to other environmental stress stimuli, including response to antimicrobial agents. Therefore, understanding of the QS multi-layered regulatory network would help to develop alternative therapeutic intervention against *P. aeruginosa* infections. This is particularly important as antibiotic therapy is becoming increasingly ineffective due to widespread multidrug resistance. In this chapter, we will provide understanding on the four known QS systems in *P. aeruginosa*, focussing on components of each QS system, QS signal molecules structures and their known corresponding receptors. Secondly, we will show how each QS system interacts with each other and also to other stimuli to better understand the regulatory matrix in *P. aeruginosa*. Finally, we will discuss inhibitors of QS system and their potential future application.

INTRODUCTION

Bacterial communication was largely ignored until the middle of 20th century when scientists showed that pneumococci and aquatic microorganisms such as *Vibrio fisheri* and *Vibrio harveyi* (previously known as *Beneckea harveyi*) produced extracellular molecules as signals to promote their behaviour (Tomasz, 1965, Nealson et al., 1970, Greenberg et al., 1979). Classical findings on quorum sensing (QS) started in the 1970s with the observation of bioluminescence of *Vibrios* under high-density culture conditions. Strong evidence came in the 1980s with the discovery of genes required for quorum control of luminescence in *Vibrio fisheri* known as *luxI* and *luxR* which control *lux* gene transcription. While *luxI* gene encodes enzymes responsible for the production of the QS signal, N-3-oxohexanoy-L-homoserine lactone (3OC6-HSL), *luxR* encodes for a 3OC6-HSL responsive transcriptional factor of the *luxCDABEG* operon which enables bioluminescence and affects other cellular activities. These first observations on intercellular communication in bacteria via such autoinducers (AI) took several decades to be widely recognized and were termed quorum sensing (QS) in the 1990s. Since then, QS or bacterial cell-cell signaling/communicating, has been studied in multiple bacteria including many human pathogens. With its role as the responding mechanism of bacterial group behaviour in which individual cells produce small diffusible molecules (QS signals or autoinducers, AI) to communicate with other surrounding cells, the importance of QS in bacterial survival and pathogenesis via regulating a wide range of gene expression is established. QS systems nowadays are known to be involved in both co-operative and competitive interactions between microorganisms living in a cohort.

QS functions in a QS- signal dependent mode. At a sufficient concentration, QS signals cross membranes by diffusion, bind to transcriptional regulator proteins leading to the increased transcription of many genes in the bacterial genome. Mechanisms of QS systems in Gram-negative and Gram-positive bacteria are different. Gram-negative bacteria typically use fatty-acid derivatives such as N-acyl homoserine lactone (AHL) as QS signals which cross the cell membrane by diffusion without any support of cytoplasmic or cell membrane proteins. These molecules are then detected via LuxR-type receptors which are cytoplasmic transcription factors. The LuxR- type receptors consist of two functional domains: one amino-terminal domain responsible for QS signal binding and the other carboxyl-terminal domain for DNA binding. Only in the presence of QS signals, LuxR proteins fold properly resulting in typical C-terminal helix-turn-helix domains, becoming stable and being able to bind to DNA to trigger transcription (Schuster et al., 2004, Lee et al., 2006, Papenfort et al., 2016). This transduction process is thus called a one-component signal transduction system.

In Gram-positive bacteria, QS is regulated by a two-component signal transduction system. This consists of two major components, a sensor kinase protein which is localized in the membrane and a cognate response regulator protein located inside the cell. In addition, instead of using HSL/AHL chemical signals, oligopeptides are frequently used as communication molecules in Gram-positive bacteria, thus called autoinducer peptides (AIPs). These molecules are synthesized from modified and processed precursor peptides. Nowadays, signaling peptides are also found in Gram-negative bacteria. The first peptidic molecule involved in the quorum-sensing of *E. coli* acts as an interspecies extracellular death factor (EDF). It is a pentapeptide with the sequence NNWNN (Asn-Asn-Trp-Asn-Asn-OH), shown to participate in the cell death process mediated by the toxin-antitoxin system *mazEF*. EDF was also characterized in *P. aeruginosa* (Kumar et al., 2013) (Figure 3.1).

Regarding QS signals, the first identified and most intensively studied example is AHL or AI-1. The latter QS signal is AI-2 which is a furanosyl borate diester. This molecule has been characterized as a signaling molecule used in interspecies communication in Gram-negative bacteria (Figure 3.1). Over time, more QS signaling molecules and QS systems have been identified in various types of organisms including prokaryotes and eukaryotes. Although QS systems vary from species to species, each QS normally consists of an enzyme that synthesizes the signaling molecule (QS signal synthase or QS-I, *for QS inducer*) and a transcription factor that binds to the QS signal and regulates the expression of QS regulons (QS-responsive transcription factor or QS-R, *for QS receptor*) (Box 1). QS-R can upregulate QS-I leading to a phenomenon called "autoinduction" promoting synchronous gene expression of the population. Activity of QS-R depends on concentration of QS signals or autoinducers in the environment which increase proportionally with the population density. Based on the alteration in the QS signal concentration, bacteria sense changes in surrounding cell numbers and respond via the collective change of global patterns of gene expression. Notably, the processes under QS regulation are generally those that are metabolically costly and unproductive when being performed by individual cells but effective when expressed by populations of cells. Examples include such as bioluminescence, virulence factor secretion and biofilm formation.

Being an important human pathogen, *Pseudomonas aeruginosa* has been the subject of intensive investigation and is one of the models for studying QS. In the following sections, current understanding of QS in *P. aeruginosa* is presented and the potential to interfere with the QS mechanisms for human therapeutic benefit is also reviewed.

Overview on Quorum Sensing in *Pseudomonas aeruginosa*

So far, four QS systems have been indentified in *P. aeruginosa*: *las*, *iqs*, *pqs* and *rhl*. Of these, *las* and *rhl* are most extensively studied (Lee et al., 2015). They work together to allow the bacteria to respond flexibly to environmental changes, increasing their chance of survival, resisting the host response to infection and/or antimicrobial stress during treatment of infection. It is clear that QS systems play an important role in infections caused by *P. aeruginosa*. Therefore, understanding of the QS network in *P. aeruginosa* is essential for the development of therapeutic QS inhibitors (QSIs) or quorum quenching enzymes (QQEs) with the aim of modulating the bacterial virulence, antimicrobial resistance and persistence.

Las and Rhl

Elucidation of the *lux* QS system in *Vibrio* comprising *luxI* and *luxR* is the basis for the discovery of the two most prominent QS systems in *P. aeruginosa: las*, consisting of *lasI* and *lasR*, and *rhl*, consisting of *rhlI* and *rhlR*. LasR and RhlR belong to the LuxR protein family with significant homology in two distinct regions, one located in the N-terminal part of the protein required for QS signals binding, the other being in the C-terminal part required for DNA binding with a helix-turn-helix motif (Stevens et al., 1999). Like other QS-R of luxR family, LasR and RhlR work as QS-signal responsive transcription activators. (Table 3.1).

LasR, a 26 kDa protein, was initially known as a key regulator of two virulence-associated metalloproteases (the elastases *lasA* and *lasB*) and later was shown to be also required for expression of the alkaline protease *apr* and to enhance expression of exotoxin A, *toxA* (Gambello et al., 1993). Secreted proteases, particularly elastases, are important determinants of *Pseudomonas* pathogenesis. Elastases are produced by most clinical isolates and able to degrade a broad range of host cell molecules including elastin, collagen, laminin, immunoglobulins, complement components and alpha-1 antiproteinase inhibitor. ToxA, belonging to the mono-ADP-ribosyltransferase family, is the most toxic extracellular virulence factor of *P. aeruginosa*. This 68-kDa transferase irreversibly inhibits protein synthesis of host cells causing cell death (Figure 3.2).

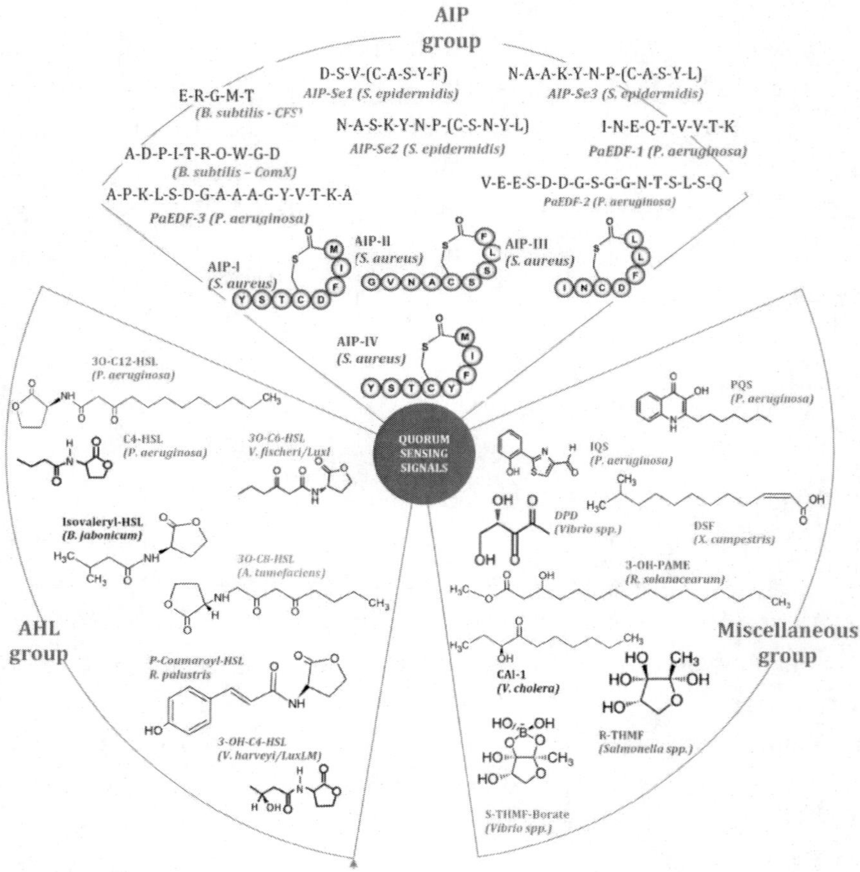

Figure 3.1. Examples of quorum sensing signals from Gram-negative and Gram-positive bacteria. (i) AHL, N-acyl homoserine lactone or AI-1, mostly found in Gram-negative bacteria, (ii) AIP, autoinducer peptide, firstly identified in Gram-positive bacteria, but also found in Gram-negative bacteria serving as interspecies communication signals (iii) Other signaling molecules including the *Vibrio* AI-2, autoinducer-2, furanosyl borate ester form, PQS, Pseudomonas quinolone signal, 2-heptyl-3-hydroxy-4(1H)-quinolone, DSF, diffusible factor, methyl dodecenoic acid, PAME, hydroxyl-palmitic acid methyl ester (*Ralstonia solanacearum*), IQS, Integrated Quorum Sensing Signal, 2-(2-hydroxyphenyl)-thiazole-4-carbaldehyde, DPD, (S)-4,5-dihydroxy-2,3-pentanedione (*Vibrio spp*), CAI-1, cholera autoinducer-1/(S)-3-hydroxytridecan-4-one, R-THMF, (2R,4S)-2-methyl-2,3,3,4-tetrahydroxytetrahydrofuran (*Salmonella* spp), S-THMF, (2S,4S)-2-methyl-2,3,3,4-tetrahydroxytetrahydrofuran (*Vibrio* spp).

Table 3.1. Cognate receptors of QS signals in *Pseudomonas aeruginosa*

Four subunits in QS system	Signal molecules	Cytoplasmic receptors proteins (Ligands)	References
Las	N-(3-oxododecanoyl)-L-homoserine lactone (3-oxo-C12-HSL)	LasR, QscR	*Schuster* et al., *2004*, *Lee* et al., *2006*
Rhl	N-butanoyl-L-homoserine lactone (C4-HSL)	RhlR	*Winson* et al., *1995*
Pqs	2-heptyl-3-hydroxyl-4-quinolone (Pseudomonas quinolone signal)	PqsR (MvfR)	*Fletcher* et al., *2007*
Iqs	2-(2- hydroxyphenyl)-thiazole-4-carbaldehyde	Unknown	*Lee* et al., *2013*, *Papenfort* et al., *2016*

LasI, the LuxI equivalent in *P. aeruginosa*, directs the synthesis of N-(3-oxo-dodecanoyl)-homoserine lactone (3-oxo-C12-HSL) structurally related to N-3-oxohexanoy-L-homoserine lactone (3OC6-HSL) of the *lux* QS system in *Vibrio* (Figure 3.2). This QS-signal binds to and activates LasR which results in alteration of target gene transcription including the *lasI*, thus leading to autoinduction (Seed et al., 1995) (Figure 3.2). LasR also induces the expression of RsaL which is a transcriptional repressor of *lasI* producing a negative feedback loop. It is worth noting that the repression activity of RsaL is stronger than the activation activity of LasR on *lasI*, even though they do not compete for the same binding site on the promoter of *lasI*.

Similarly in the *rhl* system, RhlI promotes production of N-(butanoyl)-homoserine lactone (C4-HSL or BHL), which then interacts with the cognate response regulator RhlR, changing expression of target genes including *rhlI* resulting in an autoinduction positive feedback loop (Figure 3.2). RhlR, encoded by the rhamnolipid synthase gene cluster *rhlABR*, of which *rhlI* gene is downstream, is a key regulator of virulence genes in *P. aeruginosa*. Once RhlI is expressed, production of elastase, pyocyanin, hemolysin and rhamnolipids is also induced. Pyocyanin, a blue redox-active secondary metabolite, is the most important phenazine of *P.*

aeruginosa. It is responsible for the bluish colour of sputum and pus in *Pseudomonal* infections due to its large quantity production. It has been demonstrated to interfere with multiple cellular processes in cultured epithelial cells and mediate tissue damage and necrosis during lung infections. Rhamnolipids are amphiphilic biosurfactants consisting of rhamnose molecules linked to fatty acids. This chemical nature allows the molecules to interrupt cell membranes causing intracellular material leakage. Rhamnolipids are also known to induce biofilm disruption and inhibit adhesion of competing microorganisms.

Transcriptomic and proteomic studies reveal that nearly 10% of the *P. aeruginosa* genome (more than 300 genes) is regulated by *las* and *rhl* QS systems mostly via the binding of the complex, LasR-3-oxo-C12-HSL and/or RhlR- C4-HSL to the conserved *las- rhl* boxes located at the promoters of target genes (Schuster et al., 2007).

QscR is another cognate receptor of the *lasI*-generated 3-oxo-C12-HSL (Oinuma et al., 2011). This receptor diverges from LasR and RhlR in a number of ways. Firstly, QscR lacks an associated LuxI-type cognate synthase, thus it appears to rely on the 3-oxo-C12-HSL signal produced by *lasI* for its activity. Secondly, previous studies demonstrated that a *P. aeruginosa* mutant lacking *qscR* was hypervirulent *in vivo*, suggesting that *qscR* acts as a repressor of QS. Thirdly, QscR can bind numerous AHLs, which has led to speculation that it may play a role not only in communication within a species, but also between different species (Lee et al., 2006, Lintz et al., 2011, Ding et al., 2018). Some of the repressed QS-controlled genes are not regulated by Las or Rhl systems, while others are. QscR controls a specific regulon that overlaps with the already overlapping LasR- and RhlR-dependent regulons (Greenberg et al., 2006) (Table 3.1).

The two systems, *las* and *rhl* are strongly connected with the ability of LasI/R to regulate the transcription of *rhlI-rhlR* (Kiratisin et al., 2012). Moreover, LasI/R is also known to activate other more recently identified QS systems, *pqs* (Déziel et al., 2004) and *iqs* (Lee et al., 2013).

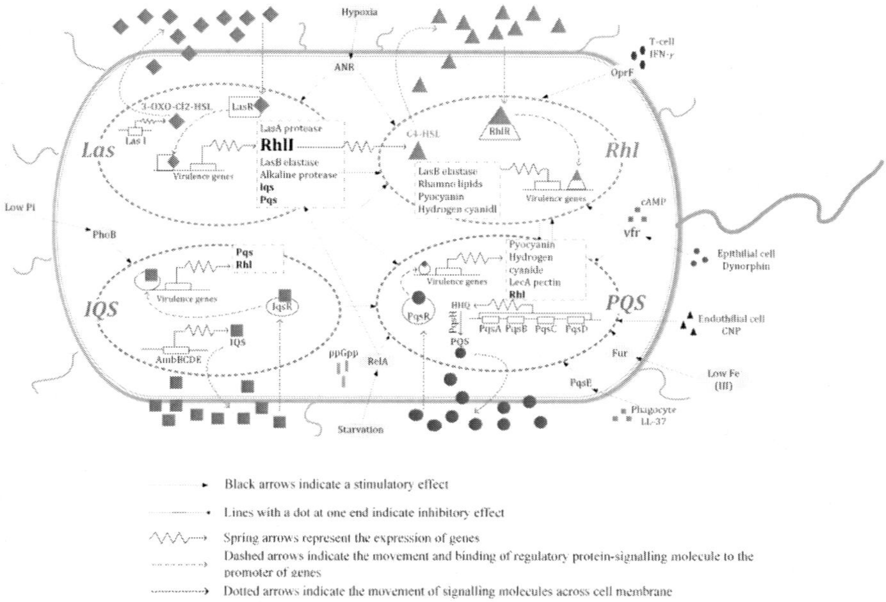

Figure 3.2. Inherent quorum sensing systems in *Pseudomonas aeruginosa.* Black arrow: stimulatory effect; Dot- ending line: inhibitory effect; Spring arrow: gene expression; Dashed arrows: movement and binding of QS signal-cognate receptors complex; Dotted arrow: movement and binding of QS signals.

Pqs and Iqs

Unlike *las* and *rhl*, *pqs* and *iqs* have distinct QS signals. The QS signal of *pqs* was characterized in 1999 and identified as 2-heptyl-3-hydroxy-4-quinolone, a kind of alkyl quinolone (AQ), called PQS for Pseudomonas quinolone signal (Pesci et al., 1999) (Figure 3.2). The biosynthesis of PQS is carried out by a gene cluster comprised of *phnAB*, *pqsABCDE* and *pqsH.*

PhnAB is an anthranilate synthase which converts the central metabolite chorismate to anthranilate. This reaction is also the first step in tryptophan biosynthesis which is also carried out by another functional anthranilate synthase, TrpEG whose function drives the pathway towards the synthesis of tryptophan rather than PQS, as in the case of PhnAB (Palmer et al., 2013). PqsA, an anthranilate-coA ligase, activates anthranilate into anthraniloyl-coA which can then be converted into 2-heptyl-4-quinolone (HHQ) by PqsB, PqsC and PqsD. HHQ is then transformed into PQS by PqsH, gene transcription of which is under the regulation of LasR (Schertzer et al., 2009). PqsR is the cognate QS transcriptional regulator of PQS. It is a LysR-type transcriptional regulator, also known as MvfR (multiple virulence factor regulator) that regulates production of multiple virulence factors including pyocyanin, elastase, PA-IL lectin and rhamnolipids. Null mutation of the *pqs* system leads to significant reduction of biofilm formation and virulence of *P. aeruginosa* on tested organisms (Cao et al., 2001, Gallagher et al., 2002). Furthermore, PqsR-PQS also binds to the promoter region of the *pqsABCDE* and *phnAB* operons which encode enzymes for the synthesis of PQS as mentioned above. This provides a positive feedback loop as in the case of *las* and *rhl* to promote its own activity and synthesis of more than 50 other low-molecular weight compounds which are involved in regulating transcription of genes related to virulence and infection (Kitao et al., 2018). PqsR-PQS also activates RhlR/I which in turn suppresses *pqs* operons (Figure 3.2). PQS is produced mainly during the late stationary phase, inducing expression of both *lasB* and *rhlI.*

Like PQS, the *iqs* signal is structurally unrelated to previously known signaling molecules and identified as 2-(2-hydroxyphenyl)-thiazole-4-carbaldehyde, called IQS for Integrated quorum sensing signal. The biosynthesis of IQS is determined by a non-ribosomal peptide synthase gene cluster, *ambBCDE.* This process is tightly regulated by *las* under normal culture conditions but activated by phosphate depletion which frequently occurs during host infection. It was shown that PhoB, the phosphate starvation response protein can activate IQS production, which in turn induces *pqs* then *rhl* expression, resulting in a by-pass mechanism to trigger virulence gene expression without LasR activity.

Recently, it has been reported that under stressful conditions, *P. aeruginosa* produces QS peptides mediating interspecies bacterial cell death, called extracellular death factor (EDF) (Kumar et al., 2013).

Three EDFs have been identified in *P. aeruginosa* including one nonapeptide and two hexadecapeptides named, *Pa*EDF-1, -2, -3 respectively. The EDF have been shown to trigger *mazEF*- and *ydcDE*-mediated cell death in *E. coli* and *B. subtilis*, respectively. New understanding on QS of *P. aeruginosa* is still accumulating.

The QS in Action

All the four QS systems mentioned above interact with each other helping individual *P. aeruginosa* in a population to respond smoothly to environmental changes. Many QS-regulated products are secreted as exoproteases and can be used by all individuals in the cohort, as so called "public goods". Each individual has to pay a metabolic cost for the synthesis of "public goods" which is beneficial for the whole cohort, including itself. Various studies have used *P. aeruginosa* as a model to study population behavior. For example, *P. aeruginosa* uses QS to control the production of elastase which is required for bacterial growth on media containing casein as the sole source of carbon and energy. The cooperation is also observed in rhamnolipid production in biofilm models and virulence factor production in infection models. The cooperation is maintained not only by bringing beneficial effects but also by sanctioning "cheaters", QS-mutants taking advantage of "public goods", in the population. For instance, besides the production of elastase as a public good, QS of *P. aeruginosa* also regulates "private goods" like periplasmic enzymes responsible for adenosine catabolism that gave an advantage to QS-wild type but not QS-mutant bacteria. Furthermore, "cheaters" can also be "punished" via toxic effects of factors produced by "cooperators". For example, QS of *P. aeruginosa* can induce both hydrogen cyanide production and cyanide resistance thus limiting the growth of QS- mutants inhibited by hydrogen cyanide. Additionally, production of costly "public goods" is delayed via QS thus suppressing emergence of "cheaters". Rhamnolipids, which are needed for swarming, are controlled via QS as well as by availability of carbon and nitrogen. They are not secreted constitutively but at stationary phase where production cost is low (Abisado et al., 2018).

QS- regulated compounds of *P. aeruginosa* can help the bacteria compete against other bacterial species. When both *P. aeruginosa* and *Staphylococcus aureus* are co-present, *P. aeruginosa* often takes over or reduces the *S. aureus* population. *P. aeruginosa* and *S. aureus* likely compete for limiting and overlapping sources of iron which is essential for the growth of most microbial pathogens. It has been shown that AI-2 produced by staphylococci upregulates several major virulence genes of *P. aeruginosa* including extracellular protease (lasA/B), rhamnosyltransferase involved in rhamnolipid synthesis (rhlA), exotoxins (exoT, exoS, exoY) and phenazines (phzA1 and phzA2. LasA, for example, was shown to exhibit staphylolytic protease activity, thus called staphylolysin. This enzyme enhances elastolytic activity via the cleavage of Gly-Gly bonds which are abundant in elastin and degrades the pentaglycine cross-linkage in the staphylococcal cell wall peptidoglycan, thereby rapidly lyses staphylococci even at low concentrations (Kessler et al., 1993). In addition, rhamnolipids are known to induce Gram-positive cell lysis via cell membrane intercalation. Another QS-related compound, 4-hydroxy-2-heptylquinoline N-oxide (HQNO) and the QS-regulated product, pyocyanin suppress the growth of *S. aureus* by inhibiting oxidative respiration. Interestingly, this suppression activity also induces the emergence of *S. aureus* small colony variants (SCV), thereby increasing the persistence of *S. aureus* (Hotterbeeks et al., 2017). *In vivo*, it is frequently observed that infections of *P. aeruginosa* develop after *S. aureus* infections, indicating the dominance of *P. aeruginosa* over *S. aureus*; and the co-infections are generally more severe and take longer to respond to treatment as the result of virulence gene upregulation and induction of SCVs (Hotterbeeks et al., 2017).

The QS-regulated products are also known to be important in the competition of *P. aeruginosa* over other *Proteobacteria* species (Abisado et al., 2018). However, the QS systems of *P. aeruginosa* could also be exploited by other bacteria. *Vibrio cholerae*, for example, was demonstrated to be able to suppress QS systems of *P. aeruginosa* using it strongest QS signal, CAI-1 at quite low concentration and even inhibit the growth of *P. aeruginosa* at high concentrations. It is likely that CAI-1 disturbs QS receptor organization/orientation within the membrane of *P. aeruginosa* at low concentration and induces membrane pore formation at high concentration resulting in cell lysis (Ganin et al., 2012).

Furthermore, QS is also known to regulate bacteriophage defense mechanisms. Generally, QS-mediated resistance provides protection during high risk of phage infection such as in case of high cell density. The relationship between QS and bacteriophage sensitivity in *P. aeruginosa* was first observed in 1999 (Glessner et al., 1999). It was shown recently that the expression of CRISPR-Cas (clustered regularly interspaced short palindromic repeat and CRISPR-associated proteins) genes which are responsible for the antiviral bacterial adaptive immune system is significantly down-regulated in *P. aeruginosa* with *lasI* and *rhlI* deletion (Høyland-Kroghsbo et al., 2017).

QS plays an important role during infection. In amoeba infection models, QS-deficient *P. aeruginosa* show markedly decreased virulence (Pukatzki et al., 2002). Data from mouse models show that *las* and *rhl* are needed for *P. aeruginosa* to disseminate, causing pathology and mortality (Smith et al., 2003B). In chronic infections such as cystic fibrosis, *P. aeruginosa* grows and proliferates within the highly viscous sputum in the lungs of patients. Data showed that in sputum samples from cystic fibrosis patients, HSLs were detected and QS-regulated genes expressed. These result in reduction of swimming motility and premature activation of the Pseudomonas quinolone signal cell-to-cell signaling cascade in *P. aeruginosa* in cystic fibrosis sputum. Furthermore, HSLs are also known to interfere with the human immune system by suppressing production of cytokines including IL-12 and TNF-alpha, thereby suppressing the immune response (Smith et al., 2003B). In the late stage of chronic infection, QS systems are reduced suggesting adaptation of the bacteria to the lung of cystic fibrosis patients (Bjarnsholt et al., 2009).

In the presence of antibiotic stress, QS systems appear to play a very important role in protecting *P. aeruginosa*. Most data indicated that QS can induce biofilm formation, covering *P. aeruginosa* cells in extracellular matrix. The biofilm formation process involves in the gene expression change leading to the downregulation of polar flagella, upregulation of Type IV pili followed by the overproduction of extracellular matrix. In *P. aeruginosa*, polysaccharides Pel, Psl and alginate are important in biofilm structure. In addition, proteins, lipids, biosurfactants and particularly extracellular DNA (eDNA) also contribute greatly to biofilm maturation. Recently, the QS system *pqs* has been shown to trigger programmed cell death in *P. aeruginosa* which releases extracellular DNA, promoting biofilm formation and antibiotic tolerance for the rest of the population (Hazan et al., 2016). Experimental results showed that biofilm cells are highly resistant to antibiotics, up to 1000 fold in comparison to free swimming, planktonic ones. The exact mechanism of the antibiotic resistance of biofilm is not yet fully understood. It may result from a number of factors including: reduction in antibiotic diffusion and uptake due to biofilm structure and presence of anionic eDNA; higher concentration of extracellular antibiotic-degrading enzymes trapped within the biofilm matrix; low metabolic rate of cells deep inside the biofilm and horizontal gene transfer between bacterial strains and species living together in the biofilm. All of these mechanisms seem related to QS and complex array of adaptive gene expression changes.

QS can also induce other extracellular enzymes such as peroxidases which protect *P. aeruginosa* from reactive oxygen species, e.g., hydrogen peroxide and also promotes tolerance to the carbapenem antibiotic, meropenem (Maura et al., 2016). Furthermore, interference with QS can change the susceptibility of *P. aeruginosa* to antibiotics. For examples, addition of a QS signal, acyl-homoserine lactone 3-OC12-HSL to *P. aeruginosa* cultures exposed to carbenicillin and ciprofloxacin increase the number of persister cells (Möker et al., 2010.). In contrast, adding compounds that can block QS in *P. aeruginosa* such as furanone can significantly impair *P. aeruginosa* growth and increase antibiotic efficacy. We will discuss in details in the next part of this chapter on the potential of inhibiting QS.

Overall, QS is particularly important for *P. aeruginosa* to sense, communicate and coordinate various behaviors. Over the past few decades, a significant knowledge of QS has been built, providing the opportunity for development of novel approaches to treat infection. Progress in interference with QS systems of *P. aeruginosa* is reviewed in the next sections. However, further understanding on QS systems and functions is essential to find efficient ways to interfere, thereby generating successful alternative treatment in pseudomonal infections.

INHIBITION OF QUORUM SENSING

Inhibition of QS is called quorum quenching (QQ) which was first described in 2000. An enzyme from *Bacillus* called lactonase AiiA for autoinducer inactivator A was found with the ability to inactivate N-acyl HSL of *Erwinia carotova*. The production of lactonase AiiA enzyme resulted in a significant reduction of extracellular pectolytic enzyme activities, and attenuation of pathogenicity on potato, eggplant, Chinese cabbage, carrot, celery, cauliflower, and tobacco (Dong et al., 2000). This finding stimulates researchers to overcome disease by targeting the QS system using QQ molecules. It is an interesting and promising approach whereby the specific interference with the QS would reduce the pathogen virulence but not their viability. Thus, development of antibiotic resistance to conventional antibiotics would be less likely to trigger. To date, several QS interruption approaches have been explored including: (i) targeting QS signals, (ii) targeting the cognate receptors and (iii) targeting other potential regulators. The agents are summarized in Figure 3.3 and discussed in details as below.

Targeting QS Signals: Degradation, Blocking and Interference with the Synthesis

Degradation of QS Signals- The QQ Enzymes

As mentioned above, QQ enzymes such as lactonases have potential to inhibit QS system via degradation of QS signals like HSLs. The role of lactonases was previously demonstrated when Reimmann and his colleagues cloned lactonases encoding gene into pME6863 vector (Reimmann et al., 2002). Expression of the gene by *P. aeruginosa* resulted in significantly reduction of quorum sensing signals such as virulence gene expression and swarming motility. So far, identified QQ enzymes mainly target HSLs and are classified into: (i) HSL-lactonase that cleaves the ester bond in the HSL ring moiety, (ii) HSL-acylases that hydrolyzes the amide bond between the acyl chain and HSL and (iii) HSL-oxidoreductases that modify the 3-oxo-substituents of the HSLs. Previously, these enzymes were not considered suitable for therapeutic usage due to the large size of the enzyme molecules and difficulty in delivering active enzymes to the site of infection. Recently, however, QQ enzymes, particularly HSL-lactonases, have been tested for their efficacy in mammalian models of pulmonary infection. For example, lactonase AiiM showed activity against *P. aeruginosa* infection in an acute pneumonia mouse model (Migiyama et al., 2013). Inhalation of a versatile lactonase, SsoPox, originally known as paraoxonase (Pox) from the hyperthermophilic *Sulfolobus solfactircus* (Sso) reduced *P. aeruginosa* QS resulting in lower mortality in a rat pneumonia model (Hraiech et al., 2014). In addition, lactonase used with ciprofloxacin was demonstrated to prevent the infection against *P. aeruginosa* and reduce mortality in a murine burn mouse model (Gupta et al., 2015). Another enzyme known as PvdQ acylase is able to hydrolyse AHL signaling molecules irreversibly. Following that, *P. aeruginosa* QS system was inhibited by the enzyme when tested *in vitro* and in infection model of *Caenorhabditis elegans*. To date, application of PvdQ intranasally in mouse pulmonary infection model resulted in significant decrease of *P. aeruginosa* virulence (Rémy et al., 2018). In a different study, SsoPox were conjugated natural biopolymers, alginate and chitosan to improve the enzyme delivery without changing their activity (Rhoads et al. 2018). Some of the QQ enzymes have already been coated onto medical devices to reduce nosocomial infections including SSoPox on coatings, acylases on catheters and nanofibers and lactonase in topical creams (Rémy et al., 2018). (Table 3.2)

Table 3.2. Current application of quorum quenching agents

Applied field	QQ agent	Application	References
Medicine	Azithromycin Lactonase enzyme PvdQ	Treatment of ventilator-associated pneumonia Prevent infection against *P. aeruginosa* in a murine burn wound model Candidate for antibacterial therapy against *Pseudomonas* infections	*Van Delden et al., 2012* *Gupta* et al., *2015* *Evelina* et al., *2009*
Aquaculture	Crude aqueous extracts of *Proteus mirabilis* Lactonases and acylases from genus *Erythrobacter*, *Labrenzia*, *Bacterioplanes*	Inhibit biofilm formation by decreasing the production of virulence factors Inhibit biofilm formation by degrading 3OXOC12-AHLs	*Yu et al., 2018* *Rehman et al., 2018*
Environment	Piper betle extract (PBE)	Anti-biofouling in membrane Bioreactors (MBRs)	*Muhammad* et al., *2012*
Medical devices	QSI 5-fluorouracil, Furanones, DHP TZD-8 Acylase from *Aspergillus melleus* FS3, RNA-III inhibiting peptide (RIP) analog (YAPWTNF-NH2). Lactonase from Bacillus sp. ZA12 PLL SsoPox from *S. Solfataricus*	Coated on central nervous catheter Coated on catheters Coated on urinary catheters Coated on silicon catheter Coated on prosthesis Topical treatment Immobilized onto polyurethane coating and nanoalumina membranes	*Jacobsen* et al., *2008*, *Walz* et al., *2010* *Ozcelik* et al., *2017* *Shenderovich* et al., *2015* *Ivanova* et al., *2015b* *Grover* et al., *2016* *Cirioni* et al., *2013* *Gupta* et al., *2015* *Ng* et al., *2011*, *Hraiech* et al., *2014*, *Guendouze* et al., *2017*

Blocking QS Signals

Another approach to interrupt the secreted QS signals is to block them using QS signal-binding molecules. For example, cyclodextrins, the cyclic oligosaccharides and their derivatives were reported to bind HSLs, form inclusion complexes with them and disrupt QS signaling. These compounds were synthesized and tested on different Gram-negative bacteria including *P. aeruginosa* and the synthetic alkylamine-modified cyclodextrins had strongest inhibitory effects on QS (Rémy et al., 2018).

QS signal-binding molecules of particular research interest are anti QS-signal antibodies. For instance, a monoclonal antibody called RS2-1G9 generated against a 3-oxo-dodecanoyl HSL analogue hapten demonstrated potent inhibitory activity of 3-oxo-C12-HSL-based QS in *P. aeruginosa in vitro*. Interestingly, this antibody, when tested *in vivo* using murine model, can sequester 3-oxo-C12-HSL resulting in diminished level of this QS signal in the serum and tissue, preventing the activation of cellular stress kinase pathways, protecting 3-oxo-C12-HSL-mediated apoptosis in murine macrophages (Rémy et al., 2018). Furthermore, a panel of sheep-mouse chimeric monoclonal antibodies have been generated and they, particularly HSL-2 and HSL-4 antibodies, have shown significant effect on increasing survival rate of worms and mice challenged with *P. aeruginosa* (Palliyil et al., 2013). Similar results were also obtained in other bacteria such as *S. aureus*. With the use of anti-autoinducer monoclonal antibody, AP4-24H11, researchers were able to inhibit QS *in vitro*, suppress *S. aureus* pathogenicity *in vivo* and provide complete protection against a lethal *S. aureus* challenge (Rémy et al., 2018). In short, it is suggested that QS signals do not only regulate gene expression in *P. aeruginosa* but also contribute to exert cytotoxicity on hosts, thus anti-QS signal antibodies and related molecules are highly promising immunopharmacotherapy for the treatment of *P. aeruginosa* and other bacterial infections.

Interference of QS Signal Biosynthesis

One further possible option targeting QS signals is to interfere with QS signal biosynthesis. The synthase enzymes LasI and RhlI depend on the amino acid and fatty acid biosynthetic pathways to synthesize HSL. The HSL ring is derived from S- adenosyl methionine (SAM) of the amino acid biosynthetic pathway and acyl chains from the acyl-carrier proteins (acyl-ACP) of the fatty acid biosynthetic pathway. Any alterations in these two pathways can affect the activity of LasI and RhlI. An inhibitor screening study on LasI and RhlI synthases revealed that salicylic acid, tannic acid and trans-cinnamaldehyde can be used to inhibit LasI and RhlI (Chang et al., 2014). Alternatively, the biosynthesis of PQS requires anthranilate whose production can be inhibited by the methyl anthranilate, an anthranilate analogue (Calfee et al., 2001). Such small molecules with the ability to inhibit QS constitute a group called QS inhibitors or QSIs which can also be used in targeting the cognate receptors such as LasR and RhlR as discussed in the next section.

Targeting the Cognate Receptors: LasR, RhlR and PqsR (MvfR) Inactivation

QS Signal Analogues

To date, most QSIs studies have focussed on inhibiting the transcriptional regulators of the QS system, i.e., compounds are tested to interfere with the binding of signaling molecules to the cognate receptor such as LasR, RhlR or PqsR (MvfR) through competitive inhibition. In order to inactivate LasR and RhlR, HSL analogues have been used as antagonists for 3-oxo-C12-HSL and C4-HSL to compete for binding sites on LasR and RhlR.

In 1996, a series of structural analogs of 3-oxo-C12-HSL were synthesised and tested for their ability to act as autoinducers and antagonists of LasR. Results showed that the length of the acyl side chain and the presence of 3-oxo moiety of HSL are crucial for antagonist activity (Passador et al. 1996). The obtained data facilitated our understanding on the interaction between AI and LasR and thus promoted rational design of therapeutic QSIs. In addition, structural comparison between 3-oxo-C12-HSL and C4-HSL revealed important epitopes needed for activation and inhibition of LasR activity. In 2003, a group of compounds sharing a common aniline-ring structure was identified to compete with 3-oxo-C12-HS and inhibit activation of LasR and elastase production (Smith et al., 2003A). Recently, meta-bromo-thiolactone (mBTL) was demonstrated to represses LasR and RhlR, blocking the formation of biofilms and production of virulence factor. Furthermore, mBTL also showed protective activity in *Caenorhabditis elegans and human lung epithelial cell infection models against the QS-mediated killing effect of P. aeruginosa* (O'Loughlin et al., 2013).

For inhibiting PqsR (MvfR), homologues of PQS or alkyl-quinolone analogues have been synthesized and evaluated (Ilangovan et al., 2013). Study showed that PQS homologues/quinazolinone analogues bind to the ligand binding pocket of PqsR in a similar orientation like the native agonist QS signals but result in antagonist effect. These binding significantly decrease virulence gene expression and biofilm development of *P. aeruginosa*.

Furanones

Another QSI with high potential, furanone has also been extensively studied. In the early 2000s, using GeneChip® microarray technology, a research group showed that synthetic derivatives of natural furanone compounds can specifically inhibit QS systems and QS-induced virulence gene expression (Hentzer et al., 2003). Furanone compounds like C-30 were shown to exhibit higher antagonistic activity to LasR than RhlR. The finding of these potential furanone compounds originated from previous studies attempting to attenuate bacterial pathogenesis using natural signal antagonists derived from marine organisms. Scientists had observed that seaweeds, although lacking advanced immune systems, produce secondary metabolites to protect against colonizing organisms. Biologically active halogenated furanones, for examples are found abundantly in seaweeds and these secondary metabolites are released and inhibit the colonization of both prokaryotes and eukaryotes (de Nys et al., 1995). Nowadays, much attention has been paid to screening novel compounds with potential to specifically target QS systems thus reducing *P. aeruginosa* virulence and pathogenicity without affecting bacterial growth and viability.

Application of furanones increased bacterial susceptibility to tobramycin and sodium dodecyl sulphate (SDS) in a *P. aeruginosa* biofilm model and suppressed QS, promoting immune clearance of the infecting bacteria in a mouse pulmonary infection model (Hentzer et al., 2003). These data are exciting, because they show that QSIs are effective in *in vivo* models promising potential treatment in patients suffering from *P. aeruginosa* infections. Furthermore, in this study, microarray analysis showed that exposure to halogenated furanone compounds resulted in repression of 80% of QS- regulated genes indicating the specificity of the compounds.

Terrein

Recently, inhibitor targeting *P. aeruginosa* virulence factors and biofilms known as terrein was isolated from *Aspergillus terreus*. This five-membered- ketone ring compound is more stable than lactone-skeleton compounds such as furanones. Terrein also demonstrated an excellent inhibition activity on biofilm formation and production of elastase, pyocyanin and rhamnolipids without imposing a selective survival pressure on *P. aeruginosa* in both *in vitro* setting and *in vivo* (murine airway infection model. The expression level of QS-regulatory genes (*lasI*, *lasR*, *rhlI*, *rhlR*, *pqsA*, *phnB*, *pqsH*, and *pqsR)* and key QS-controlled genes (*lasA*, *lasB*, *aprE* encoding for zinc- protease transporter/alkaline protease transporter, *rhlA* and *phzA2* being located in the gene cluster *phzRABCDEFG* and encoding for phenazine biosynthesis protein A) was affected by terrein dose-dependently (Kim et al., 2018). Interestingly, terrein was also identified as the inhibitor of c-di-GMP signaling. It was demonstrated that terein decreases the activity of a diguanylate cyclase (DGC), thus reducing the intracellular 3,5-cyclic diguanylic acid (c-di-GMP) levels. These findings suggested an inherent connection between QS signaling and c-di-GMP metabolism in *P. aeruginosa.*

Phenolic Compounds: Coumarins and Related Compounds

Among plant secondary metabolites, phenolic compounds with antimicrobial activity have potential to be exploited as QSIs. Coumarins, for example, are a large family of fragrant compounds containing fused benzene and 2 pyrone rings, (benzopyrone) which can be considered as a subclass of lactones. Some coumarins are phytoalexins which are synthesized by plant tissues to protect plants from pathogenic infections while other coumarins secreted by bacteria and fungi could also serve as antibiotics such as the aminocoumarins: novobiocin (albamycin/ cathomycin) and coumermycin. The first study to report coumarins as QSIs was a docking analysis of different compounds from the Traditional Chinese Medicine Library on QS transcriptional activator TraR of *Agrobacterium tumefaciens* (Zeng et al., 2008). It showed that the hydroxycoumarins, esculatine and esculin were able to bind to the TraR signal binding site.

Furthermore, coumarins also suppressed expression of *PqsA* and *rhlI* and several key genes involved in *las* and *iqs* systems in the pathogen (Zhang et al. 2018). They inhibited biofilm formation (particularly esculetin, umbelliferone and coumarin), phenazine production, motility and production of other virulence factors such as protease and pyocyanin (Reen et al. 2018, Zhang et al., 2018). The protective effect of coumarin against *P. aeruginosa* was also demonstrated in an *in vitro* wound model and a *Lucilia sericata* infection model. Moreover, similar to terrein, courmarin also changed the expression of genes related to cyclic diguanylate (c-di-GMP) metabolism resulting in a significant reduction of cellular c-di-GMP level in *P. aeruginosa* (Zhang et al., 2018). Two fucocoumarins isolated from grapefruit juice, bergamottin and dihydroxybergamottin were shown to inhibit biofilm formation in *P. aeruginosa*. Both fucocoumarins suppressed the HSL-QS signaling pathway (Girennavar et al., 2008). The synthetic coumarins such as courmarin- benzimidazone hybrids generated recently showing broad antibacterial spectrum also extend the possibility of using this QSI group (Reen et al., 2018).

In brief, coumarin and its related derivatives are universal QSIs with ability to affect not only *P. aeruginosa* but also other Gram-negative and Gram-positive bacteria. They are highly potential and require further investigation for future clinical application. The plant food extracts and phytochemical compounds with potential QSIs are summarized in Table 3.3.

Table 3.3. Natural plants, herbs, spices and other compounds as QSI in *Pseudomonas aeruginosa*

Source	Antagonist	Biological action	References
Plant derived extracts			
Capparis spinose extract		Inhibit biofilm formation, EPS production and swarming motility	*Isaac-Abraham* et al., *2011*
Grapefruit	Furocoumarins	Inhibit autoinducer signaling and biofilm formation	*Girennavar* et al., *2008*
Nutmeg extract	Malabaricone C	Inhibit both the lasR and rhlR systems	*Chong* et al., *2011*
Broccoli and other cruciferous vegetable extracts	Sulforaphane Erucin	Inhibit LasA and LasB protease activities	*Ganin* et al., *2013*
Sweet basil (Ocimum basilicum L.)	Rosmarinic acid	Decrease the expression of the elastase and protease, as well as biofilm formation	*Walker* et al., *2004*
Banana		Inhibit swarming motility	*Fatima* et al., *2010*
Raspberry (*Rubus idaeus* L.), blueberry (*Vaccinium angustifolium Aiton*),		Inhibite AHL activity-mediated violacein production	*Kalia., 2013*
Edible fruits: *Ananas comosus* (L.) Merr. (pineapple) and *Manilkara zapota* (Sapodilla)		QSI against pyocyanin pigment, staphylolytic protease, elastase production, and biofilm formation abilities	*Musthafa* et al., *2010*
Lilium brownie extracts Panax pseudoginseng extracts		Inhibit biofilm formation	*Yeo* et al., *2012*
Ficus carica and *Perilla frutescens*		Inhibit swarming motility	*Sun* et al., *2014*
Phyllanthus amarus methanolic extract		Inhibit swarming motility, pyocyanin production and lecA expression	*Priya* et al., *2013*
Fresh herbs and spices extract			
Clove extracts and ginseng		Inhibit swarming motility, pyocyanin production and LecA gene expression	*Wu* et al., *2011*, *Krishnan* et al., *2012*
Garlic extract	Disulphides and trisulphides	Inhibit LuxR-based QSI	*Rasmussen* et al., *2005*
Syzygium aromaticum extract		Reduce pyocyanin synthesis	*Krishnan* et al., *2012*
Cinnamon oil and eugenol		Inhibit pyocyanin, protease and elastase production	*Zhou* et al., *2013*, *Kalia* et al., *2015*

Table 3.3. (Continued)

Source	Antagonist	Biological action	References
Curcuma longa L. (turmeric)	Curcurmin	Reduce elastase and protease production	*Rudrappa and Bais., 2008*
Others			
Catechin		Reduction of pyocyanin production, elastase activity and biofilm formation, Reduction of lasB, lasI, lasR, rhlA, rhlI, rhlR gene expression	*Vandeputte* et al., *2010*
Ajoene		Reduce rhlA and lasB promoter activity Reduce AHL levels and rhamnolipid production	*Jakobsen* et al., *2018* *Galloway* et al., *2012*
Caffeine (1, 3, 7-trimethylxanthine)		Inhibit AHL production and swarming	*Norizan* et al., *2013*
Trans-anethole		Reduces virulence factor production but also inhibits swarming motility	*Zhou* et al., *2013*, *Kalia* et al., *2015*
Dihydroxybergamottin, bergamottin and Rosmarinic acid		Inhibit biofilm formation	*Walker* et al., *2004*
Eriodictyol, Taxifolin, Phenylacetic acid, Naringenin Malabaricone C Farnesol		Reduce pyocyanin production and elastase activity, inhibit biofilm formation	*Vandeputte* et al., *2011*, *Cugini* et al., *2010*, *Chong* et al., *2011*, *Musthafa* et al., *2012*, *Vandeputte* et al., *2011*
Iberin		Reduce Rhl and Lux reporters, rhamnolipid production, lasB and rhlA gene expression	*Jakobsen* et al., *2012a*
Salicylic acid		Reduce AHL production, twitching and swimming motility	*Bandara* et al., *2006*

Other Miscellaneous Synthetic QSIs

In 2016, Nizalapur and his colleagues designed and evaluated the QSI activity of synthesized compounds of novel N-arylisation-based glycosamine derivatives. These compounds are generated via ring-opening reactions of N-acylisatins. It was observed that these synthetic molecules exhibited inhibitory effects on QS of both *P. aeruginosa* and *E. coli.* They also resulted in a reduction of pyocyanin and biofilm production (Nizalapur et al., 2017).

Quite recently, a group of PqsR synthetic inhibitors structurally distinct from PQS, so called non-ligand-based benzamine-benzimidazole (BB) was identified using whole-cell high through-put screening and structure-activity relationship (SAR) analysis. These molecules bind PqsR, inhibiting the PqsR regulon in multi-drug resistant isolates, suppressing *P. aeruginosa* infection in mice without affecting bacterial growth and viability (Starkey et al., 2014, Kitao et al., 2018). Furthermore, they are the first synthetic compounds identified to reduce the formation of antibiotic-tolerant persister cells. These molecules have potential for developing a new generation of therapeutic drugs for treatment of *P. aeruginosa* and other bacterial infections.

Besides, there are also other studies identifying QSIs with significant bacterial inhibitory activity using biosensors. However, this procedure is prone to bias thus, researchers should be really careful when accessing whether what they found is really QSI (Defoirdt et al., 2013).

In brief, synthetic QSI molecules could potentially be used as alternatives to antibiotics in the future and contribute to overcome antimicrobial resistance or used for the study of QS for treatment of pathogen infections. At current, multiple compounds with unrelated structures have been developed from screening studies. Some compounds can be antagonist of one cognate receptor but agonist of another, emphasizing the inherent complexity of QS systems and difficulty in the development of successful QSIs.

Targeting Other Potential Regulators

Whilst QS systems regulate a large number of genes, they themselves are also controlled by others. Data have shown that GacA, Vfr and RelA can upregulate *lasR* expression (Figure 3.2) (Smith et al., 2003B). Deletion of GacA leads to decreased production of virulence factors and reduced pathogenicity of *P. aeruginosa* in burn-mouse model infection. Vfr knockout also resulted in a similar observation with LasR depression and reduction of virulence gene expression (Smith et al., 2003B). These findings indicate that there are other proteins that could serve as QS inhibition targets.

CONCLUSION

QS controls many traits of the bacteria, enabling them to response effectively to any change in the surroundings. It directs activities that are beneficial when performed by groups of bacteria sharing one shelter or environment. It also regulates many bioactivities such as competence development, nutrient usage, bioluminescence, sporulation, antibiotic production, motility and biofilm formation. For clinically-related bacteria such as *P. aeruginosa*, QS has been shown to have great impact on its pathogenesis via controlling virulence factor production and secretion which can either increase its virulence, persistence, resistance or host modulation ability.

To date, with the decline in number of newly introduced antibiotics and a rapid increase of antimicrobial resistance, the approach of finding new drugs to reduce the pathogenicity and virulence of the pathogens and/or increase the effectivity of the current antibiotics would be highly attractive. QSIs are not only highly potential in inhibiting bacteria thus themselves can be regarded as antibiotics but also able to disrupt biofilm formation, increasing efficacy of other antibiotics. QQ approach also bring promising results on the host side with significant improvement of clinical outcome in *in vivo* models. It is no doubt that research on QS and QQ will bring us impressive understanding on microbial world and great benefit on various aspects including medicine, pharmacy and even ecology in the future.

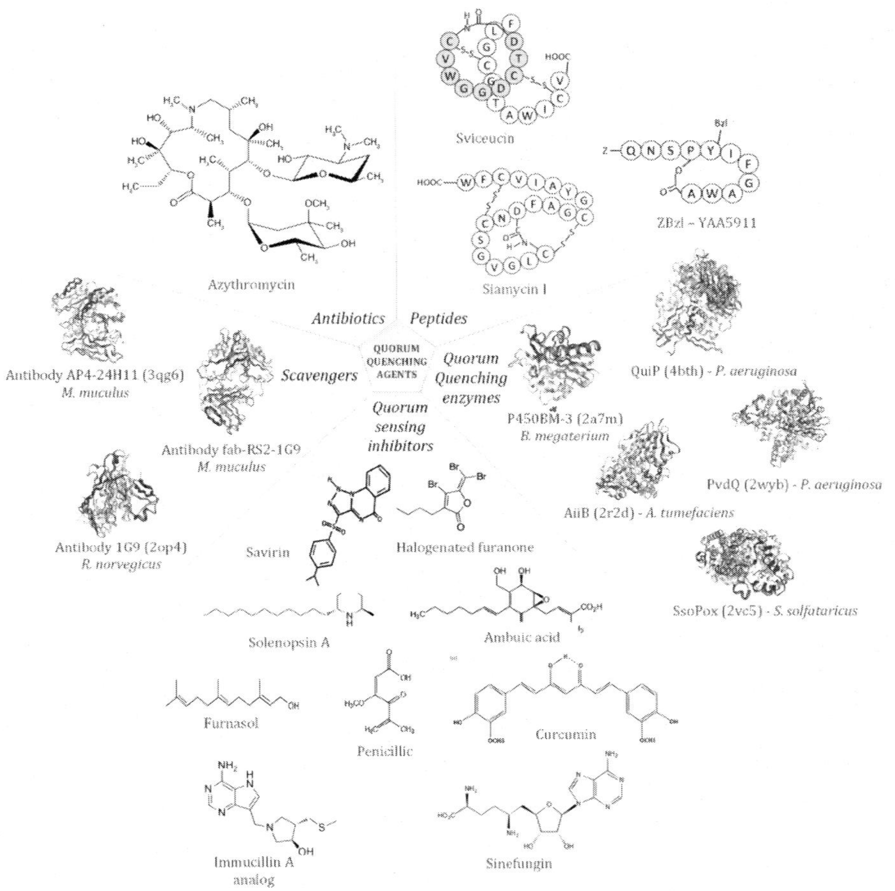

Figure 3.3. Quorum quenching agents. (A) Quorum sensing inhibitors, (B) Quorum quenching enzymes, (C) QS signal scavengers, (D) Other QS-suppressors.

ACKNOWLEDGMENT

This work is partially supported by Vietnam National Foundation for Science and Technology Development (Nafosted) under grant number 108.04-2018.08.

REFERENCES

Abisado, R. G., Benomar, S., Klaus, J. R., Dandekar, A. A. & Chandler, J. R. (2018). Bacterial Quorum Sensing and Microbial Community Interactions. *mBio.*, *9*(3), pp. e02331-17.

Bandara, M. B., Zhu, H., Sankaridurg, P. R. & Willcox, M. D. (2006). Salicylic acid reduces the production of several potential virulence factors of *Pseudomonas aeruginosa* associated with microbial keratitis. *Investigative Ophthalmology & Visual Science*, *47*(10), pp. 4453-4460.

Bjarnsholt, T., Jensen, P. Ø., Fiandaca, M. J., Pedersen, J., Hansen, C. R., Andersen, C. B., Pressler, T., Givskov, M. & Høiby, N. (2009). *Pseudomonas aeruginosa* biofilms in the respiratory tract of cystic fibrosis patients. *Pediatric Pulmonology*, *44*(6), pp. 547-558.

Calfee, M. W., Coleman, J. P. & Pesci, E. C. (2001). Interference with *Pseudomonas* quinolone signal synthesis inhibits virulence factor expression by *Pseudomonas aeruginosa*. *Proceedings of the National Academy of Sciences*, *98*(20), pp. 11633-11637.

Cao, H., Krishnan, G., Goumnerov, B., Tsongalis, J., Tompkins, R. & Rahme, L. G. (2001). A quorum sensing-associated virulence gene of *Pseudomonas aeruginosa* encodes a LysR-like transcription regulator with a unique self-regulatory mechanism. *Proceedings of the National Academy of Sciences*, *98*(25), pp. 14613-14618.

Chang, C. Y., Krishnan, T., Wang, H., Chen, Y., Yin, W. F., Chong, Y. M., Tan, L. Y., Chong, T. M. & Chan, K. G. (2014). Non-antibiotic quorum sensing inhibitors acting against N-acyl homoserine lactone synthase as druggable target. *Scientific Reports*, *4*, p. 7245.

Chong, Y. M., Yin, W. F., Ho, C. Y., Mustafa, M. R., Hadi, A. H. A., Awang, K., Narrima, P., Koh, C. L., Appleton, D. R. & Chan, K. G. (2011). Malabaricone C from *Myristica cinnamomea* exhibits anti-quorum sensing activity. *Journal of Natural Products*, *74*(10), pp. 2261-2264.

Cirioni, O., Mocchegiani, F., Cacciatore, I., Vecchiet, J., Silvestri, C., Baldassarre, L., Ucciferri, C., Orsetti, E., Castelli, P., Provinciali, M. & Vivarelli, M. (2013). Quorum sensing inhibitor FS3-coated vascular graft enhances daptomycin efficacy in a rat model of staphylococcal infection. *Peptides*, *40*, pp. 77-81.

Cugini, C., Morales, D. K. & Hogan, D. A. (2010). Candida albicans-produced farnesol stimulates *Pseudomonas* quinolone signal production in LasR-defective *Pseudomonas aeruginosa* strains. *Microbiology*, *156*, p. 3096.

De Nys, R., Steinberg, P. D., Willemsen, P., Dworjanyn, S. A., Gabelish, C. L. & King, R. J. (1995). Broad spectrum effects of secondary metabolites from the red alga *Delisea pulchra* in antifouling assays. *Biofouling*, *8*(4), pp. 259-271.

Déziel, E., Lépine, F., Milot, S., He, J., Mindrinos, M. N., Tompkins, R. G. & Rahme, L. G. (2004). Analysis of *Pseudomonas aeruginosa* 4-hydroxy-2-alkylquinolines (HAQs) reveals a role for 4-hydroxy-2-heptylquinoline in cell-to-cell communication. *Proceedings of the National Academy of Sciences*, *101*(5), pp. 1339-1344.

Ding, F., Oinuma, K. I., Smalley, N. E., Schaefer, A. L., Hamwy, O., Greenberg, E. P. & Dandekar, A. A. (2018). The *Pseudomonas aeruginosa* orphan quorum sensing signal receptor QSCR regulates global quorum sensing gene expression by activating a single linked operon. *MBio*, *9*(4), pp. e01274-18.

Dong, Y. H., Xu, J. L., Li, X. Z. & Zhang, L. H. (2000). AiiA, an enzyme that inactivates the acylhomoserine lactone quorum-sensing signal and attenuates the virulence of *Erwinia carotovora*. *Proceedings of the National Academy of Sciences*, *97*(7), pp. 3526-3531.

Evelina, P., Mariana, W., NadalJimenez, P., Gudrun, K., Rita, S. & Wim, J. (2009). Quorum-Quenching Acylase Reduces the Virulence of *Pseudomonas aeruginosa* in a *Caenorhabditis elegans. Infection Model Antimicrobial Agents and Chemotherapy.*, *53*(11), pp. 4891-4897.

Fatima, Q., Zahin, M., Khan, M. S. A. & Ahmad, I. (2010). Modulation of quorum sensing controlled behaviour of bacteria by growing seedling, seed and seedling extracts of leguminous plants. *Indian Journal Of Microbiology*, *50*(2), pp. 238-242.

Fletcher, M. P., Diggle, S. P., Crusz, S. A., Chhabra, S. R., Cámara, M. & Williams, P. (2007). A dual biosensor for 2-alkyl-4-quinolone quorum-sensing signal molecules. *Environmental Microbiology*, *9*(11), pp. 2683-2693.

Gallagher, L. A., McKnight, S. L., Kuznetsova, M. S., Pesci, E. C. & Manoil, C. (2002). Functions required for extracellular quinolone signaling by *Pseudomonas aeruginosa*. *Journal of Bacteriology*, *184*(23), pp. 6472-6480.

Gambello, M. J., Kaye, S. U. S. A. N. & Iglewski, B. H. (1993). LasR of *Pseudomonas aeruginosa* is a transcriptional activator of the alkaline protease gene (apr) and an enhancer of exotoxin A expression. *Infection and Immunity*, *61*(4), pp. 1180-1184.

Ganin, H., Rayo, J., Amara, N., Levy, N., Krief, P. & Meijler, M. M. (2013). Sulforaphane and erucin, natural isothiocyanates from broccoli, inhibit bacterial quorum sensing. *Med Chem Comm*, *4*(1), pp. 175-179.

Girennavar, B., Cepeda, M. L., Soni, K. A., Vikram, A., Jesudhasan, P., Jayaprakasha, G. K., Pillai, S. D. & Patil, B. S. (2008). Grapefruit juice and its furocoumarins inhibits autoinducer signaling and biofilm formation in bacteria. *International Journal of Food Microbiology*, *125*(2), pp. 204-208.

Glessner, A., Smith, R. S., Iglewski, B. H. & Robinson, J. B. (1999). Roles of *Pseudomonas aeruginosa* las and rhl quorum-sensing systems in control of twitching motility. *Journal of Bacteriology*, *181*(5), pp. 1623-1629.

Greenberg, E. P., Hastings, J. W. & Ulitzur, S. (1979). Induction of luciferase synthesis in *Beneckea harveyi* by other marine bacteria. *Archives of Microbiology*, *120*(2), pp. 87-91.

Grover, N., Plaks, J. G., Summers, S. R., Chado, G. R., Schurr, M. J. & Kaar, J. L. (2016). Acylase-containing polyurethane coatings with anti-biofilm activity. *Biotechnology and Bioengineering*, *113*(12), pp. 2535-2543.

Guendouze, A., Plener, L., Bzdrenga, J., Jacquet, P., Rémy, B., Elias, M., Lavigne, J. P., Daudé, D. & Chabrière, E. (2017). Effect of quorum quenching lactonase in clinical isolates of *Pseudomonas aeruginosa* and comparison with quorum sensing inhibitors. *Frontiers in Microbiology*, *8*, p. 227.

Gupta, P., Chhibber, S. & Harjai, K. (2015). Efficacy of purified lactonase and ciprofloxacin in preventing systemic spread of *Pseudomonas aeruginosa* in murine burn wound model. *Burns*, *41*(1), pp. 153-162.

Hazan, R., Que, Y. A., Maura, D., Strobel, B., Majcherczyk, P. A., Hopper, L. R., Wilbur, D. J., Hreha, T. N., Barquera, B. & Rahme, L. G. (2016). Auto poisoning of the respiratory chain by a quorum-sensing-regulated molecule favors biofilm formation and antibiotic tolerance. *Current Biology*, *26*(2), pp. 195-206.

Hentzer, M., Wu, H., Andersen, J. B., Riedel, K., Rasmussen, T. B., Bagge, N., Kumar, N., Schembri, M. A., Song, Z., Kristoffersen, P. & Manefield, M. (2003). Attenuation of *Pseudomonas aeruginosa* virulence by quorum sensing inhibitors. *The EMBO Journal*, *22*(15), pp. 3803-3815.

Høyland-Kroghsbo, N. M., Paczkowski, J., Mukherjee, S., Broniewski, J., Westra, E., Bondy-Denomy, J. & Bassler, B. L. (2017). Quorum sensing controls the *Pseudomonas aeruginosa* CRISPR-Cas adaptive immune system. *Proceedings of the National Academy of Sciences*, *114*(1), pp. 131-135.

Hraiech, S., Hiblot, J., Lafleur, J., Lepidi, H., Papazian, L., Rolain, J. M., Raoult, D., Elias, M., Silby, M. W., Bzdrenga, J. & Bregeon, F. (2014). Inhaled lactonase reduces *Pseudomonas aeruginosa* quorum sensing and mortality in rat pneumonia. *PLoS One*, *9*(10), p. e107125.

Ilangovan, A., Fletcher, M., Rampioni, G., Pustelny, C., Rumbaugh, K., Heeb, S., Cámara, M., Truman, A., Chhabra, S. R., Emsley, J. & Williams, P. (2013). Structural basis for native agonist and synthetic inhibitor recognition by the *Pseudomonas aeruginosa* quorum sensing regulator PqsR (MvfR). *PLoS Pathogens*, *9*(7), p. e1003508.

Issac-Abraham, S. V. P., Palani, A., Ramaswamy, B. R., Shunmugiah, K. P. & Arumugam, V. R. (2011). Antiquorum sensing and antibiofilm potential of *Capparis spinosa*. *Archives of Medical Research*, *42*(8), pp. 658-668.

Ivanova, K., Fernandes, M. M., Francesko, A., Mendoza, E., Guezguez, J., Burnet, M. & Tzanov, T. (2015a). Quorum-quenching and matrix-degrading enzymes in multilayer coatings synergistically prevent bacterial biofilm formation on urinary catheters. *ACS Applied Materials & Interfaces*, *7*(49), pp. 27066-27077.

Ivanova, K., Fernandes, M. M., Mendoza, E. & Tzanov, T. ((2015b). Enzyme multilayer coatings inhibit *Pseudomonas aeruginosa* biofilm formation on urinary catheters. *Applied Microbiology and Biotechnology*, *99*(10), pp. 4373-4385.

Jacobsen, S. Á., Stickler, D. J., Mobley, H. L. T. & Shirtliff, M. E. (2008). Complicated catheter-associated urinary tract infections due to *Escherichia coli* and *Proteus mirabilis*. *Clinical Microbiology Reviews*, *21*(1), pp. 26-59.

Jakobsen, T. H., Bragason, S. K., Phipps, R. K., Christensen, L. D., van Gennip, M., Alhede, M., Skindersoe, M., Larsen, T. O., Høiby, N., Bjarnsholt, T. & Givskov, M. (2012). Food as a source for quorum sensing inhibitors: iberin from horseradish revealed as a quorum sensing inhibitor of *Pseudomonas aeruginosa*. *Applied Environmental Microbiology*, *78*(7), pp. 2410-2421.

Jakobsen, T. H., Warming, A. N., Vejborg, R. M., Moscoso, J. A., Stegger, M., Lorenzen, F., Rybtke, M., Andersen, J. B., Petersen, R., Andersen, P. S. & Nielsen, T. E. (2017). A broad range quorum sensing inhibitor working through sRNA inhibition. *Scientific Reports*, *7*(1), pp. 9857.

Kalia, M., Yadav, V. K., Singh, P. K., Sharma, D., Pandey, H., Narvi, S. S. & Agarwal, V. (2015). Effect of cinnamon oil on quorum sensing-controlled virulence factors and biofilm formation in *Pseudomonas aeruginosa*. *PLoS One*, *10*(8), pp. e0135495.

Kim, B., Park, J. S., Choi, H. Y., Yoon, S. S. & Kim, W. G. (2018). Terrein is an inhibitor of quorum sensing and c-di-GMP in *Pseudomonas aeruginosa*: a connection between quorum sensing and c-di-GMP. *Scientific Reports*, *8*(1), pp. 8617.

Kitao, T., Lepine, F., Babloudi, S., Walte, F., Steinbacher, S., Maskos, K., Blaesse, M., Negri, M., Pucci, M., Zahler, B. & Felici, A. (2018). Molecular insights into function and competitive inhibition of

Pseudomonas aeruginosa multiple virulence factor regulator. *MBio*, *9*(1), pp. e02158-17.

Krishnan, T., Yin, W. F. & Chan, K. G. (2012). Inhibition of quorum sensing-controlled virulence factor production in *Pseudomonas aeruginosa* PAO1 by Ayurveda spice clove (*Syzygium aromaticum*) bud extract. *Sensors*, *12*(4), pp. 4016-4030.

Kumar, S., Kolodkin-Gal, I. & Engelberg-Kulka, H. (2013). Novel quorum-sensing peptides mediating interspecies bacterial cell death. *MBio*, *4*(3), pp. e00314-13.

Lee, J. & Zhang, L. (2015). The hierarchy quorum sensing network in *Pseudomonas aeruginosa*. *Protein & Cell*, *6*(1), pp. 26-41.

Lee, J., Wu, J., Deng, Y., Wang, J., Wang, C., Wang, J., Chang, C., Dong, Y., Williams, P. & Zhang, L. H. (2013). A cell-cell communication signal integrates quorum sensing and stress response. *Nature Chemical Biology*, *9*(5), p. 339.

Lee, J. H., Lequette, Y. & Greenberg, E. P. (2006). Activity of purified QscR, a *Pseudomonas aeruginosa* orphan quorum-sensing transcription factor. *Molecular Microbiology*, *59*(2), pp. 602-609.

Lintz, M. J., Oinuma, K. I., Wysoczynski, C. L., Greenberg, E. P. & Churchill, M. E. (2011). Crystal structure of QscR, a *Pseudomonas aeruginosa* quorum sensing signal receptor. *Proceedings of the National Academy of Sciences*, *108*(38), pp. 15763-15768.

Marques, C., Davies, D. & Sauer, K. (2015). Control of biofilms with the fatty acid signaling molecule cis-2-decenoic acid. *Pharmaceuticals*, *8*(4), pp. 816-835.

Maura, D., Hazan, R., Kitao, T., Ballok, A. E. & Rahme, L. G. (2016). Evidence for direct control of virulence and defense gene circuits by the *Pseudomonas aeruginosa* quorum sensing regulator, MvfR. *Scientific Reports*, *6*, pp. 34083.

Migiyama, Y., Kaneko, Y., Yanagihara, K., Morohoshi, T., Morinaga, Y., Nakamura, S., Miyazaki, T., Hasegawa, H., Izumikawa, K., Kakeya, H. & Kohrogi, H. (2013). Efficacy of AiiM, an N-acylhomoserine lactonase, against *Pseudomonas aeruginosa* in a mouse model of acute pneumonia. *Antimicrobial Agents and Chemotherapy*, *57*(8), pp. 3653-3658.

Möker, N., Dean, C. R. & Tao, J. (2010). *Pseudomonas aeruginosa* increases formation of multidrug-tolerant persister cells in response to quorum-sensing signaling molecules. *Journal of Bacteriology*, *192*(7), pp. 1946-1955.

Musthafa, K. S., Ravi, A. V., Annapoorani, A., Packiavathy, I. S. V. & Pandian, S. K. (2010). Evaluation of anti-quorum-sensing activity of edible plants and fruits through inhibition of the N-acyl-homoserine lactone system in *Chromobacterium violaceum* and *Pseudomonas aeruginosa*. *Chemotherapy*, *56*(4), pp. 333-339.

Musthafa, K. S., Sivamaruthi, B. S., Pandian, S. K. & Ravi, A. V. (2012). Quorum sensing inhibition in *Pseudomonas aeruginosa* PAO1 by antagonistic compound phenylacetic acid. *Current Microbiology*, *65*(5), pp. 475-480.

Nealson, K. H., Platt, T. & Hastings, J. W. (1970). Cellular control of the synthesis and activity of the bacterial luminescent system. *Journal of Bacteriology*, *104*(1), pp. 313-322.

Ng, F. S., Wright, D. M. & Seah, S. Y. (2011). Characterization of a phosphotriesterase-like lactonase from Sulfolobus solfataricus and its immobilization for disruption of quorum sensing. *Applied Environmental Microbiology*, *77*(4), pp. 1181-1186.

Norizan, S., Yin, W. F. & Chan, K. G. (2013). Caffeine as a potential quorum sensing inhibitor. *Sensors*, *13*(4), pp. 5117-5129.

O'Loughlin, C. T., Miller, L. C., Siryaporn, A., Drescher, K., Semmelhack, M. F. & Bassler, B. L. (2013). A quorum-sensing inhibitor blocks *Pseudomonas aeruginosa* virulence and biofilm formation. *Proceedings of the National Academy of Sciences*, *110*(44), pp. 17981-17986.

Oinuma, K. I. & Greenberg, E. P. (2011). Acyl-homoserine lactone binding to and stability of the orphan *Pseudomonas aeruginosa* quorum-sensing signal receptor QscR. *Journal of Bacteriology*, *193*(2), pp. 421-428.

Ozcelik, B., Ho, K. K. K., Glattauer, V., Willcox, M., Kumar, N. & Thissen, H. (2016). Poly (ethylene glycol)-based coatings combining low-biofouling and quorum-sensing inhibiting properties to reduce bacterial colonization. *ACS Biomaterials Science & Engineering*, *3*(1), pp. 78-87.

Palmer, G. C., Jorth, P. A. & Whiteley, M. (2013). The role of two *Pseudomonas aeruginosa* anthranilate synthases in tryptophan and quorum signal production. *Microbiology*, *159*, p. 959.

Papenfort, K. & Bassler, B. L. (2016). Quorum sensing signal–response systems in Gram-negative bacteria. *Nature Reviews Microbiology*, *14*(9), p. 576.

Passador, L., Tucker, K. D., Guertin, K. R., Journet, M. P., Kende, A. S. & Iglewski, B. H. (1996). Functional analysis of the *Pseudomonas aeruginosa* autoinducer PAI. *Journal of Bacteriology*, *178*(20), pp. 5995-6000.

Pesci, E. C., Milbank, J. B., Pearson, J. P., McKnight, S., Kende, A. S., Greenberg, E. P. & Iglewski, B. H. (1999). Quinolone signaling in the cell-to-cell communication system of *Pseudomonas aeruginosa*. *Proceedings of the National Academy of Sciences*, *96*(20), pp. 11229-11234.

Priya, K., Yin, W. F. & Chan, K. G. (2013). Anti-quorum sensing activity of the traditional Chinese herb, *Phyllanthus amarus*. *Sensors*, *13*(11), pp. 14558-14569.

Pukatzki, S., Kessin, R. H. & Mekalanos, J. J. (2002). The human pathogen *Pseudomonas aeruginosa* utilizes conserved virulence pathways to infect the social amoeba *Dictyostelium discoideum*. *Proceedings of the National Academy of Sciences*, *99*(5), pp. 3159-3164.

Rasmussen, T. B., Skindersoe, M. E., Bjarnsholt, T., Phipps, R. K., Christensen, K. B., Jensen, P. O., Andersen, J. B., Koch, B., Larsen, T. O., Hentzer, M. & Eberl, L. (2005). Identity and effects of quorum-sensing inhibitors produced by *Penicillium* species. *Microbiology*, *151*(5), pp. 1325-1340.

Reen, F. J., Gutiérrez-Barranquero, J. A. & Parages, M. L. (2018). Coumarin: a novel player in microbial quorum sensing and biofilm formation inhibition. *Applied Microbiology and Biotechnology*, *102*(5), pp. 2063-2073.

Rehman, Z. U. & Leiknes, T. (2018). Quorum-quenching bacteria isolated from Red Sea sediments reduce biofilm formation by *Pseudomonas aeruginosa*. *Frontiers in Microbiology*, *9*, pp. 1354.

Reimmann, C., Ginet, N., Michel, L., Keel, C., Michaux, P., Krishnapillai, V., Zala, M., Heurlier, K., Triandafillu, K., Harms, H. & Défago, G. (2002). Genetically programmed autoinducer destruction reduces virulence gene expression and swarming motility in *Pseudomonas aeruginosa* PAO1. *Microbiology*, *148*(4), pp. 923-932.

Rémy, B., Mion, S., Plener, L., Elias, M., Chabrière, E. & Daudé, D. (2018). Interference in bacterial quorum sensing: A biopharmaceutical perspective. *Frontiers in Pharmacology*, *9*, pp. 203.

Rhoads, M., Hauk, P., Gupta, V., Bookstaver, M., Stephens, K., Payne, G. & Bentley, W. (2018). Modification and Assembly of a Versatile Lactonase for Bacterial Quorum Quenching. *Molecules*, *23*(2), pp. 341.

Rudrappa, T. & Bais, H. P. (2008). Curcumin, a known phenolic from Curcuma longa, attenuates the virulence of *Pseudomonas aeruginosa* PAO1 in whole plant and animal pathogenicity models. *Journal of Agricultural and Food Chemistry*, *56*(6), pp. 1955-1962.

Schertzer, J. W., Boulette, M. L. & Whiteley, M. (2009). More than a signal: non-signaling properties of quorum sensing molecules. *Trends in Microbiology*, *17*(5), pp. 189-195.

Schuster, M. & Greenberg, E. P. (2007). Early activation of quorum sensing in *Pseudomonas aeruginosa* reveals the architecture of a complex regulon. *BMC Genomics*, *8*(1), pp. 287.

Schuster, M., Urbanowski, M. L. & Greenberg, E. P. (2004). Promoter specificity in *Pseudomonas aeruginosa* quorum sensing revealed by DNA binding of purified LasR. *Proceedings of the National Academy of Sciences*, *101*(45), pp. 15833-15839.

Seed, P. C., Passador, L. & Iglewski, B. H. (1995). Activation of the *Pseudomonas aeruginosa* lasI gene by LasR and the *Pseudomonas* autoinducer PAI: an autoinduction regulatory hierarchy. *Journal of Bacteriology*, *177*(3), pp. 654-659.

Shenderovich, J., Feldman, M., Kirmayer, D., Al-Quntar, A., Steinberg, D., Lavy, E. & Friedman, M. (2015). Local sustained-release delivery systems of the antibiofilm agent thiazolidinedione-8 for prevention of catheter-associated urinary tract infections. *International Journal of Pharmaceutics*, *485*(1-2), pp. 164-170.

Smith, K. M., Bu, Y. & Suga, H. (2003A). Library screening for synthetic agonists and antagonists of a *Pseudomonas aeruginosa* autoinducer. *Chemistry & Biology*, *10*(6), pp. 563-571.

Smith, R. S. & Iglewski, B. H. (2003B). *Pseudomonas aeruginosa* quorum sensing as a potential antimicrobial target. *The Journal of Clinical Investigation*, *112*(10), pp. 1460-1465.

Starkey, M., Lepine, F., Maura, D., Bandyopadhaya, A., Lesic, B., He, J., Kitao, T., Righi, V., Milot, S., Tzika, A. & Rahme, L. (2014). Identification of anti-virulence compounds that disrupt quorum-sensing regulated acute and persistent pathogenicity. *PLoS Pathogens*, *10*(8), pp. e1004321.

Stevens, A. M. & Greenberg, E. B. (1999). Transcriptional activation by LuxR. In Dunny, G. M. and Winans, S. C. *Cell-cell signaling in bacteria*. ASM Press.

Sun, S., Li, H., Zhou, W., Liu, A. & Zhu, H. (2014). Bacterial quorum sensing inhibition activity of the traditional Chinese herbs, *Ficus carica* L. and *Perilla frutescens*. *Chemotherapy*, *60*(5-6), pp. 379-383.

Tomasz, A. (1965). Control of the competent state in *Pneumococcus* by a hormone-like cell product: an example for a new type of regulatory mechanism in bacteria. *Nature*, *208*(5006), pp. 155.

Van Delden, C., Köhler, T., Brunner-Ferber, F., François, B., Carlet, J. & Pechère, J. C. (2012). Azithromycin to prevent *Pseudomonas aeruginosa* ventilator-associated pneumonia by inhibition of quorum sensing: a randomized controlled trial. *Intensive Care Medicine*, *38*(7), pp. 1118-1125.

Vandeputte, O. M., Kiendrebeogo, M., Rajaonson, S., Diallo, B., Mol, A., El Jaziri, M. & Baucher, M. (2010). Identification of catechin as one of the flavonoids from *Combretum albiflorum* bark extract that reduces the production of quorum-sensing-controlled virulence factors in *Pseudomonas aeruginosa* PAO1. *Applied Environmental Microbiology*, *76*(1), pp. 243-253.

Vandeputte, O. M., Kiendrebeogo, M., Rasamiravaka, T., Stevigny, C., Duez, P., Rajaonson, S., Diallo, B., Mol, A., Baucher, M. & El Jaziri, M. (2011). The flavanone naringenin reduces the production of quorum sensing-controlled virulence factors in *Pseudomonas aeruginosa* PAO1. *Microbiology*, *157*(7), pp. 2120-2132.

Walker, T. S., Bais, H. P., Déziel, E., Schweizer, H. P., Rahme, L. G., Fall, R. & Vivanco, J. M. (2004). *Pseudomonas aeruginosa*-plant root interactions. Pathogenicity, biofilm formation, and root exudation. *Plant Physiology*, *134*(1), pp. 320-331.

Walz, J. M., Avelar, R. L., Longtine, K. J., Carter, K. L., Mermel, L. A. & Heard, S. O. (2010). Anti-infective external coating of central venous catheters: a randomized, noninferiority trial comparing 5-fluorouracil with chlorhexidine/silver sulfadiazine in preventing catheter colonization. *Critical Care Medicine*, *38*(11), pp. 2095-2102.

Winson, M. K., Camara, M., Latifi, A., Foglino, M., Chhabra, S. R., Daykin, M., Bally, M., Chapon, V., Salmond, G. P. & Bycroft, B. W. (1995). Multiple N-acyl-L-homoserine lactone signal molecules regulate production of virulence determinants and secondary metabolites in *Pseudomonas aeruginosa. Proceedings of the National Academy of Sciences*, *92*(20), pp. 9427-9431.

Wu, H., Lee, B., Yang, L., Wang, H., Givskov, M., Molin, S., Høiby, N. & Song, Z. (2011). Effects of ginseng on *Pseudomonas aeruginosa* motility and biofilm formation. *FEMS Immunology & Medical Microbiology*, *62*(1), pp. 49-56.

Yeo, S. S. M. & Tham, F. Y. (2012). Anti-quorum sensing and antimicrobial activities of some traditional Chinese medicinal plants commonly used in South-East Asia. *Malaysian Journal of Microbiology*, *8*(1), pp. 11-20.

Yu, S., Zhu, X., Zhou, J. & Cai, Z. (2018). Biofilm inhibition and pathogenicity attenuation in bacteria by *Proteus mirabilis*. *Royal Society Open Science*, *5*(4), pp. 170702.

Zeng, Z., Qian, L., Cao, L., Tan, H., Huang, Y., Xue, X., Shen, Y. & Zhou, S. (2008). Virtual screening for novel quorum sensing inhibitors to eradicate biofilm formation of *Pseudomonas aeruginosa*. *Applied Microbiology and Biotechnology*, *79*(1), pp. 119.

Zhang, Y., Sass, A., Van Acker, H., Wille, J., Verhasselt, B., Van Nieuwerburgh, F., Kaever, V., Crabbé, A. & Coenye, T. (2018). Coumarin reduces virulence and biofilm formation in *Pseudomonas aeruginosa* by affecting quorum sensing, type III secretion and c-di-GMP levels. *Frontiers in Microbiology*, *9*, pp. 1952.

Zhou, L., Zheng, H., Tang, Y., Yu, W. & Gong, Q. (2013). Eugenol inhibits quorum sensing at sub-inhibitory concentrations. *Biotechnology Letters*, *35*(4), pp. 631-637.

In: *Pseudomonas aeruginosa* ISBN: 978-1-53616-100-7
Editors: C. Fah Wong and H. Idris

Chapter 4

HEAVY METALS BIOSORPTION WITH *Pseudomonas aeruginosa*

***Kuan Seng How*[1] *and Sylvia Chieng*[2]**
[1]Department of Mechanical and Material Engineering, Lee Kong Chian Faculty of Engineering and Science, University Tunku Abdul Rahman, Sg Long Campus, Selangor, Malaysia
[2]Centre for Biotechnology and Functional Food, Faculty of Science and Technology, Universiti Kebangsaan Malaysia, Malaysia

ABSTRACT

Heavy metals can enter our environment through various pathways, most of which end up in our water and soil. Modern industries are responsible to a large degree for such pollution of the environment. Many industrial activities, anthropogenic in nature such as mining operations, metallurgical industry, metal plating industry, fertilizer industry, battery manufacturing, tannery operations, and chemical manufacturing involve the discharge of heavy metals into water sources. Heavy metal pollution does not only affect the quality of water bodies, it also jeopardizes the health of human beings and animals by being one of the major contaminants in food supply. Conventional remediation techniques to remove these elements such as chemical precipitation, ion-exchange, electrowinning, and reverse osmosis typically involve high costs and large environmental footprints. Bioremediation is a process that uses microorganisms to transform harmful substances to nontoxic compounds through various mechanisms including biosorption, bioaccumulation and bioreduction. In particular, biosorption in recent years has been identified as a potential bioremediation method to replace conventional methods. Biosorbents that can be used to remove metal ions in wastewater are algae, fungi, bacteria and other polysaccharide materials. *Pseudomonas aeruginosa* is one example of bacteria that can reduce heavy metals content through the process of biosorption and has been shown to be effective for major toxic metals such as Cr, Cd, Pb, Al, Co, Cu, Hg, Mn, Ni and Zn among others. This review examines the mechanisms employed by *P. aeruginosa* in the biosorption process of the specific metals or metal groups of interest, the latest research to date in heavy metals biosorption using the bacteria, challenges in scaling-up to industrial applications and future perspectives on heavy metal biosorption in wastewater systems.

INTRODUCTION

Rapid industrial development has led to massive discharge of heavy metal wastewaters into the environment, particularly in developing countries. These pollutants come from industries such as metal plating facilities, mining operations, fertilizer industries, pulp and paper industries, pesticide manufacturing and batteries manufacturing. Unlike organic contaminants, heavy metals are non-biodegradable and accumulate in living organisms and thus, pose a health risk to humans. Toxic heavy metals that are present in industrial wastewaters include chromium, lead, cadmium, mercury, nickel, copper and zinc. Of these metals, mercury, lead, cadmium and chromium (VI) are considered as toxic while the rest such as copper, nickel, cobalt and zinc pose a lower risk but are nevertheless harmful when used extensively and in large amounts beyond the limits set by environmental authorities. The last few decades have seen stricter controls and regulations to mitigate the effects of these contaminants both in developing and developed countries (Vijayaraghavan and Yun, 2008; Abdel-Aty et al. 2013; Abdi and Kazemi, 2015).

Various techniques have been employed to treat industrial effluents bearing heavy metals. Precipitation of metals is the most common method to remove heavy metals from water due to low costs and simplicity in operation (Ku and Jung, 2001; Fu and Wang, 2011). Coagulants such as lime, iron salts and organic polymers are added to turn soluble metal ions into insoluble precipitates. The process typically leads to the generation of large amounts of toxic sludge which requires costly disposal techniques. The process also lacks selectivity when a series of metal ions are present in the solution and is rather ineffective for low metal concentrations. Ion-exchange processes have also been widely implemented in wastewater treatment due to their numerous advantages over other processes such as high throughput capacity, high efficiency and rapid kinetics (Kang et al., 2004). The ion-exchange resin used in the process has the specific ability to exchange its cations with the metal ions in solution treated. Electrowinning is another technique used to remove heavy metals from effluents (Ahalya et al., 2003; Abdi and Kazemi, 2015). The process is widely used in mining and metallurgical operations as well as electrical and electronics industry. Typical metals found in effluents such as Cd, Co, Ag, Ni, Pb and Zn are recovered through electro-deposition using insoluble anodes. Reverse osmosis (RO) is a process that uses a semi-permeable membrane which allows fluid that is being purified to pass through while rejecting the undesirable contaminants. Unlike precipitation, RO is capable of removing a wide range of dissolved species from water and is becoming increasingly popular in wastewater treatment applications (Fu and Wang, 2011). Electro-coagulation is an electrochemical process which uses an electrical current to remove metals from solution. Metal ions are neutralized with ions of opposite electrical charges and precipitate as a result.

These conventional methods are typically cost-intensive and largely dependent on the metal concentrations treated. Other disadvantages include incomplete metal removal, high reagent and energy requirements and generation of large amounts of toxic sludge and other waste products (Ahalya et al., 2003). In recent years, biosorption has been identified as a potential bioremediation method to replace conventional methods.

Biosorbents that can be used to remove metal ions in wastewater are algae, fungi, bacteria, yeast and waste from the industry. *Pseudomonas aeruginosa* is one example of bacteria that can reduce heavy metals content through the process of biosorption and has been shown to be effective and versatile for major range of metals spanning from base metals and precious metals to radionuclides and rare earth elements.

BIOREMEDIATION

Bioremediation is a process that uses microorganisms to transform harmful substances to nontoxic compounds in soil, subsurface materials, water, sludge and residues (Latha and Reddy, 2013). Different detoxifying mechanisms developed by these microorganisms include the metal uptake (biosorption and bioaccumulation) and/or bioreduction (transformation) by changing the oxidative state of metal. Biosorption can be defined as a selective sequestration process of heavy metal ions or other molecules onto biological materials (Hansda et al., 2015). It is a surface phenomenon where one or more physico-chemical mechanisms take place in the metal uptake of dead or living biomass (MacHalová et al., 2015). The process involves two phases; a solid phase (sorbent or biosorbent) and a liquid phase (solvent) which contains a dissolved species to be sorbed (sorbate). The sorbent tends to have a high affinity for the sorbate and this leads to the attraction and removal of the sorbate through different mechanisms. The process continues until equilibrium is reached between the amount of sorbate bound (by the sorbent) and the amount remaining in the solution. Bioaccumulation is an intracellular metal accumulation process and involves metal binding on intracellular compounds, intracellular methylation, intracellular precipitation and other mechanisms. Unlike biosorption, the process is metabolically dependent. Bioaccumulation may be seen as a second part of the metal sequestration by living biomass (Kaduková and Virčíková, 2005; MacHalová et al., 2015).

Many researchers define biosorption as a passive process, for instance the passive uptake of metals by microbial biomass (Volesky, 1990; Malik,

2004; Gadd, 2009; Fomina and Gadd, 2014). The process can be conducted either by dead biomass or fragments of cells and tissues. However, it may also be a metabolically-independent adsorption of a sorbate through surface complexation onto cell walls and/or other outer layers being the first fast reversible adsorption step operating within a considerably slower and more complex overall bioaccumulation mechanism (Volesky, 1990; Malik, 2004; Fomina and Gadd, 2014). Bioaccumulation can be seen as a function of living organisms dependent on physical, chemical and biological mechanisms including intra- and extra-cellular processes where passive uptake plays only a limited role. Naja and Volesky (2011) distinguished passive biosorption from bioaccumulation by stressing the characteristics of the process as active, metabolically mediated and as a function of a living cell. Both mechanisms may overlap (Fomina and Gadd, 2014) and sometimes biosorption and bioaccumulation terms are used interchangeably (Ahemad and Malik, 2012).

BACTERIA AS A HEAVY METALS BIOSORBENT

Biosorption of heavy metals from aqueous solutions boast the advantages of versatility and flexibility over a wide range of applications, robustness, high efficiency in metal content reduction, low costs, selectivity of targeted metals and suitability to treat dilute heavy metal wastewaters over conventional heavy metals removal methods (Garnham, 2003; Wang and Chen, 2009; Fu and Wang, 2011). Biosorbents used in the process may range from fragile microorganisms to highly stable particles (Vijayaraghavan and Balasubramanian, 2015). For large-scale industrial applications, the biosorbent can come from: i) industrial wastes (available for free or at low cost) ii) organisms easily available in large amounts in nature iii) organisms that can be grown easily for biosorption (Park et al., 2010; Fomina and Gadd, 2014). While certain organisms may prove to be effective to remove heavy metals, certain traits render them not suitable for large-scale application. These organisms are for example pathogenic bacteria/fungi, macroalgae, macrofungi, nutritionally-fastidious extremophiles and endangered plants.

Bacteria is a major group of unicellular organisms which can be found in soil and water. They come in various shapes such as rods, spiral, cocci and filament. The suitability of bacteria as a biosorbent is mainly due to the properties and structure of the cell wall. The bacterial cell wall differs from that of other organisms in that it contains peptidoglycan. Peptidoglycan provides rigidity to the cell wall and determines the cell shape. It is also porous and serves as an impermeability barrier to small substrates (Abdi and Kazemi, 2015). Gram-positive bacteria and Gram-negative bacteria differ from each other in the sense that Gram-positive bacteria generally contain 90% peptidoglycan in the cell wall while Gram-negative bacteria contains only 10 to 20% peptidoglycan. Gram-positive bacteria comprise of a thick peptidoglycan layer connected by cross-linked amino acid bridges. Embedded in the gram-positive cell walls are also polyalcohols also known as teichoic acids, some of which are lipid responsible for the formation of lipoteichoic acids. Lipoteichoic acids link peptidoglycan to the cytoplasmic membrane because they are covalently linked to lipids within the membrane. The overall negative charge of a Gram-positive bacteria comes from the phosphodiester bonds between teichoic acid monomers. Gram-negative bacteria, on the other hand contains a much thinner cell wall compared to Gram-positive bacteria. In the cell walls lies an additional outer membrane comprised of phospholipids and lipopolysaccharides. The overall negative charge of Gram-negative bacteria comes from the highly charged lipopolysaccharide.

The use of algae as biosorbent is gaining more attention in recent years due to certain attractive features such as their low nutrient requirement, their large biomass output, and the lack of toxic substance production. Previous works have shown algae such as *Chlorella vulgaris, Spirogyra, Sargassum* sp., *Eclonia* and *Lyngbya putealis* to be capable of removing heavy metals such as Pb, Cr (VI), Cu and Ni (Aksu and Kutsal, 1991; Gupta et al., 2001; Antunes et al., 2003; Park et al., 2006; Kiran et al., 2007). Biosorbents of fungi origin on the other hand, boasts the advantage of having a high percentage of cell wall material which provides excellent metal binding properties (Horikoshi et al., 1981). Furthermore, a large amount of fungi biomass is readily available from the pharmaceutical and food industry. Among the bacteria that has been proven effective as heavy metal biosorbents are *Bacillus* sp., *Pseudomonas* sp., *Zoogloea ramigera* and *Streptomyces sp* (Strandberg et al., 1981; Norberg and Persson, 1984; Norberg and Rydin, 1984; Nakajima and Sakaguchi, 1986; Mullen et al., 1989; Addour et al., 1999). Other bacteria that have been found to be effective include *Rhodobacter sphaeroides*, *Alcaligenes eutrophus* and *Staphylococcus saprophyticus* (Seki et al., 1998; Ilhan et al., 2004). Other than algae, fungi and bacteria, yeast have also been researched for use in heavy metal removal as well as waste material from food and agricultural industries such as peat, wood, pine bark, banana pith, soybean hull, cotton hull, wool and sawdust (Das et al., 2008).

The biomass used may be living or dead as mentioned earlier. The use of dead biomass is preferred for several reasons: dead biomass typically can tolerate higher levels of toxicity, does not need certain requirements to grow in the media, can easily absorb heavy metals, involves easy regeneration and reuse and involves easier immobilisation of dead cells (Dhankhar and Hooda, 2011). However, certain living microorganisms have clear advantages over their dead counterparts, such as their ability to degrade organic pollutants and to sorb, transport, complex and transform metals. They may be used when pure biosorptive metal removal is deemed not feasible or when metabolic activity is deemed advantageous to metal removal, for instance, the use of a consortium of metal-resistant microorganisms to remove toxic metals and organic pollutants through a series of metal uptake mechanisms (Malik, 2004). The criteria for selection of a suitable biosorbent include the following factors (Das et al., 2008; Vijayaraghavan and Balasubramanian, 2015):

- Cost effective
- Capable of removing metals over a wide range of pH, temperature and physico-chemical conditions
- Good stability under extremely acidic/alkaline conditions
- Good selectivity over a range of metal ions

- No secondary pollutants released
- Easy desorption and reuse
- Little to no pretreatment required

A number species in the *Pseudomonas* genus such as *P. aeruginosa, P. fluorescens, P. cepacia, P. putida, and P. stutzeri*, have been shown to be effective in biosorption of metals (Chang et al., 1997; Pardo et al., 2003; Uslu and Tanyol, 2006; Gabr et al., 2008; Vijayaraghavan and Yun, 2008). Compared to the rest in the genus, *P. aeruginosa* and *P. putida* appears to exhibit higher metal uptake amounts spanning across different metal types. *P. aeruginosa* however, stands out in the group, with the ability to absorb a wide range of metals such as aluminum, copper, cobalt, mercury, manganese, nickel, and zinc, among others. *P. aeruginosa* is a ubiquitous Gram-negative, rod-shaped bacterium which can be isolated from different habitats. One of its preeminent characteristics is that it has a high environmental adaptability. The bacteria was used generally for the expression of rhamnolipid, biofilm research and clinical medicine research (Tang et al., 2018). Recent reports have highlighted the potential of new isolates of *P. aeruginosa* for application in bioremediation of water and soil contaminated with heavy metals (Gabr et al., 2008; Sinha and Mukherjee, 2009; Naik and Dubey, 2013; Konig-Péter et al., 2014; Muneer et al., 2016; Izrael-Živković et al., 2018). To tolerate heavy metals, *Pseudomonas* demonstrated various survival adaptations such as: energy-dependent efflux of metal ions, extracellular biosorption and cellular sequestration, bio-film formation, resistance to oxidative stress, and preservation of iron acquisition and homeostasis (Izrael-Živković et al., 2018).

MECHANISMS OF BIOSORPTION

The metal biosorption process by living cells can be defined as a two-step process where metal ions are first adsorbed to the surface of cells through interactions between metals and functional groups on the surface of the cells. The metal ions then go through the cell wall before reaching the cell membrane and cell cytoplasm. The cell wall comprises of various proteins and polysaccharides and thus allows for the binding of metal ions to active sites. Different groups of microorganisms such as algae, bacteria, cyanobacteria and fungi contain different cell wall compositions and thus, result in stark differences in in the type and amount of metal ion binding to them (Das et al., 2008). Cell walls of bacteria are principally composed of peptidoglycans consisting of linear chains of disaccharide N-acetylglucosamine-β 1, 4-N-acetylmuramic acid with peptide chains while fungal cell walls typically contain chitosan and chitin. The first step, passive biosorption is metabolism independent and proceeds rapidly by any one or a combination of the following mechanisms: ion exchange, adsorption, complexation, chelation, coordination and precipitation. In the second step, through active biosorption, metal ions penetrate the cell membrane and enter the cells (termed as intracellular binding). For the first step, it is worthwhile to note that biological material is complex and the various structural components in biomass also infer that many functional groups such as carboxyl, phosphate, hydroxyl, amino and thiol can interact with metal species to different extents and may be influenced by physico-chemical factors. The amount of metal bound is dictated by the type of biosorbent, type of metal ion present in solution, its concentrations as well as pH, solution temperature, ionic strength and ion interference by the presence of other metals.

In general, the mechanisms of heavy metal biosorption are: ion exchange, adsorption, complexation, chelation, coordination and precipitation. Ion exchange involves the interchange of ions which are formed by atoms or molecules through the gaining or losing of electrons. Adsorption is a process whereby molecules attach to solid surfaces, mostly based on electrostatic charges. There are two types of adsorption through which sequestration occurs: physical adsorption and chemisorption. Physical adsorption is non-specific and the attraction forces between molecules and solid surface are relatively weak while chemisorption is specific and involves stronger forces than those observed in physical adsorption. Metal complexation and chelation are examples of chemisorption. Complexation occurs through the formation of a species by the joining of two or more species. A metal complex is formed when one of the associating species is a metal ion. Generally, the metal atom fills the central position in any given metal complex, with adjoining anions or ligands. Chelation, on the other hand is the process of forming a chelate ring where some ligands are attached to a metal atom by more than one donor atom. Metal chelates are metal complexes where the metal atom is bound to at least two sites to an organic compound. Coordination compounds are formed when a central metal atom of a complex is bound to its immediate neighbours by covalent bonds. The covalent bonds formed are usually a result of a metal atom accepting an electron pair from each non-metal atom. Precipitation occurs when the solubility of the sorbate reaches its limit. Figure 4.1 summarises the mechanisms in graphic form.

Factors that Affect Biosorption

Temperature affects the biosorption process. For practical applications, only a narrow range of temperature differences is involved. Temperature influences biosorption to a lesser extent within a range of 20 to 35°C (Aksu et al., 1992; Veglio and Beolchini, 1997). The effect of temperature on biosorption depends on the adsorption heat (enthalpy change). For physical adsorption, adsorption reaction is exothermic and thus preferred at lower temperatures while for chemisorption, the reaction is endothermic and hence, favoured at higher temperatures (Stumm and Morgan, 1996).

Typically, a pH ranges from 3 to 6 has been found favourable for the biosorption process, due to the presence of negatively-charged carboxyl groups on the cell wall. Carboxyl groups are responsible for the extracellular binding of metal cations. At higher pH, metal complex solubility decreases and complicates the sorption process. In general, biosorption is reduced at lower pH for cationic species such as Cu, Zn, Ni and Co and increased for anionic species such as CrO^{4-} and $Au(CN)^{2-}$ (Gadd and White, 1985; Hansda et al., 2015).

Ionic strength can be defined as the competition between ions or changes in metal activity. Increased ionic strength suppresses biosorption as a result of increased electrostatic charge for cations such as Zn, Cd, Na and Cu (Schiewer and Volesky, 1997). The effect is also dependent on the manner in which the metal is bound. For example, electrostatic adsorption of Na is significantly affected by the ionic strength (Hayes and Leckie, 1987). Other sorbable ions in solution may compete with targeted metals for sorption sties (Naja and Volesky, 2011). For cationic metal ions, light metals bind less strongly than heavy metals. Schiewer and Volesky (1996) established that the overall affinity sequence follows the sequence in metal adsorption by ion exchange resin.

Biosorption Equilibrium Models

The equilibrium of the biosorption process is usually described by fitting experimental points with models typically used for the representation of isotherm adsorption equilibrium (Gadd et al., 1988; Das et al., 2008). Two widely accepted linear equilibrium adsorption isotherm models for a single solute system is given by:

Langmuir $q = \frac{q_m\, b\, C_e}{1+bC_e}$ (1)

where q is milligrams of metal accumulated per gram of biosorbent material, C_e is the metal residual concentration in solution, q_m is the maximum specific uptake related to site saturation and b is the ratio of adsorption to desorption rates.

And Freundlich $q = \pi K C_e^{1/n}$ (2)

where K and n are constants.

The models are applied at constant pH and used for the modeling in the case of one metal species present. The Langmuir isotherm assumes a finite number of uniform adsorption sites and the absence of lateral interactions between adsorbed species (Fomina and Gadd, 2014). These assumptions do not hold for more complex systems. For multi-layer adsorption, the Brunauer-Emmett-Teller (BET) isotherm is commonly used. This isotherm assumes a Langmuir equation in each layer.

$$\text{Brunauer-Emmett-Teller (BET) } q = \frac{q_m\, B\, C_e}{(C_s - C_e)\{1 + (B-1)(C_e/C_s)\}} \tag{3}$$

where C_s is the saturation concentration of the component adsorbed, B is a constant expressing the interaction energy between adsorbent and solute and q_m is a constant expressing the amount of solute adsorbed in a monolayer. While these models may provide information on metal-uptake capacity, they do not reflect any mechanisms of the sorbate uptake. Both the Freundlich and Langmuir models also do not include the effects of external environmental factors though they are capable of describing numerous biosorption isotherms in many cases (Wang and Chen, 2009).

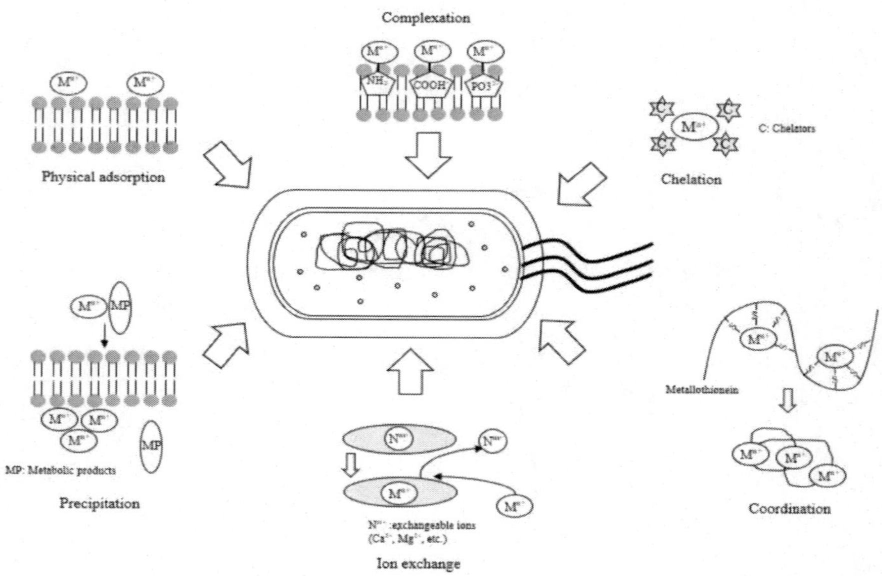

Figure 4.1. Mechanisms of heavy metal biosorption.

Pretreatment of Biosorbent

Some treatment may be performed to enhance the performance of biosorbents. Two types of treatment are common with biosorbents: chemical modification and immobilisation. Chemical modification is typically aimed at enhancing biosorption capacity and selectivity towards the targeted metal ion. Effluents generally consist of several metal ions and biosorption becomes more efficient when a biosorbent can specifically bind target ions out of the mixture. Among chemical modification procedures are pretreatment, binding site enhancement/modification and polymerization. Immobilisation is employed to create material with the right size, mechanical strength, rigidity and porosity required for metal accumulation. In certain cases, the immobilisation process may produce beads and granules that can be stripped of metals, reactivated and reused (similar to ion exchange and activated carbon processes). Immobilisation can be carried out through adsorption on inert supports, on covalent bonds in vector compounds, entrapment in polymeric matrix and cross-linking (Ahalya et al., 2003).

Despite potential advantages that come with manipulation and surface modification for biosorption, these techniques usually lead to a price increase in the biosorbent, bringing it closer to the price range of conventional ion-exchange resins and thus, compromising its low-cost advantage. Some techniques may also raise environmental, health and safety issues such as the use of hazardous chemicals, high energy electro-magnetic radiation and potentially harmful genetically modified microorganisms. While pretreatment enhances bisorbent properties in most cases, it may result in complications for the environment. The selection of the suitable pretreatment method for any given biosorbent, if necessary, requires careful consideration of these implications.

THE ROLE OF *Pseudomonas aeruginosa* IN BIOSORPTION OF HEAVY METALS

This section focuses on the interactions between our bacteria of interest i.e., *P. aeruginosa* and the specific heavy metal ions in solution. Emphasis was given on heavy metals such as chromium, cadmium, lead, aluminum, cobalt, copper, mercury, manganese, nickel and zinc which have caused considerable damage in our water systems thus far. Some discussions were also expounded on the interactions of the bacteria with radionuclides such as uranium and thorium as well as precious metals and rare earth elements. Mechanisms of biosorption are explained, as well as elaborations on the recent developments regarding the metals of interest.

Chromium

Chromium is widely used in processes such as leather tanning, electroplating, chromate preparation and metal finishing in the industry. The metal is a potential soil, surface water, ground water, sediment and air contaminant (Kamaludeen et al., 2003). The wide industrial application of chromium has led to its occurrence as a pollutant in water and soil. Some of the industries involving the use of chromium include the metallurgical, refractory, chemical and tannery operations. Though chromium can exist in oxidation states ranging from 2- to 6+, the metal commonly exists in two oxidation states, Cr(VI) and Cr(III). Cr(VI) is water-soluble, highly toxic, mutagenic and carcinogenic; while Cr(III) is less soluble, therefore less mobile, and much less toxic than Cr(VI) (Kotaś and Stasicka, 2000; Huang et al., 2009; Xu et al., 2009). Various approaches are used in industry to reduce Cr(VI) to its less toxic state Cr(III) in order to detoxify Cr(VI) polluted wastes. Conventional methods such as reverse osmosis, electrochemical process, ion exchange and adsorption on activated carbon typically consume large quantities of chemical reagents, produce large amounts of secondary toxic wastes such as sludge or require high energy demands. In recent years, biosorption, bioaccumulation and bioreduction using microorganisms has been regarded as an alternative strategy for Cr(VI) contamination since the process have been recognized to show considerably positive results (Srinath et al., 2002; Park et al., 2004; Koçberber and Dönmez, 2007; Xu et al., 2009).

Bacteria has the ability to reduce Cr(VI) under both aerobic and anaerobic conditions through electron-transport systems containing cytochromes. Under the aerobic conditions, chromate reduction is linked with soluble proteins which employ NADH as an electron donor, whereas under anaerobic conditions, chromate acts as a terminal electron acceptor through a membrane-bound reductive activity (Levinaa et al., 2007). Cr(VI) reduction may be conducted using a variety of microorganisms, including strains of *Pseudomonas sp.*, *Shewanella sp.*, *Bacillus sp.*, *Enterobacter sp.* and *Escherichia sp.*. In terms of the *Pseudomonas sp.*, removal of Cr(VI) may be conducted using *P. aeruginosa*, *P. synxantha*, *P. putida*, *P. ambigua*, *P. fluorescens*, *P. dechromaticans* and *P. chromatophila* (Suzuki et al., 1992; Lovley, 1993; Gopalan and Veeramani, 1994; McLean et al., 2000; Park et al., 2002; Cheung et al., 2006). *Pseudomonas* sp. was the first hexavalent chromium resistant strain to be isolated from waste water. Aguilera et al. (2004) investigated the resistance of *P. aeruginosa* to Cr(VI) and attributed it to the decreased uptake and/or enhanced efflux of Cr(VI) by the cell membrane (Romanenko and Koren'kov, 1977). Physiological and genetic features of *Pseudomonas* sp. render it a suitable agent for utilization in bioremediation applications. Thatheyus and Ramya (2016) observed Cr(VI) removal rate of 80% by *P. aeruginosa* and 99% by *P. fluorescens*. It was deduced that at acidic pH values, when cell wall is in direct contact with the environment, negatively charged groups are able to attract and bind to metal ions based on electrostatic forces, without cellular energy consumption which is favoured by the high surface volume ratio in bacteria (Rodríguez and Quesada, 2006). The clearest indication of typical chromium uptake mechanism by a microbial cell was provided by Jobby et al. (2018). They speculated that the Cr(VI) is first adsorbed on the microbial cell by forming chemical bonds with cell surface molecules with certain functional groups. The Cr(VI) is next reduced into Cr(III) through chromate reductase enzyme and spontaneous conversion. Cr(VI) then enters into the cell through sulphate transporters. Inside the cell, the ion is translocated with the help of specific metal binding proteins and reduced to Cr(III) through unstable oxidative states of Cr(V) and Cr(IV). Efflux of excess Cr(VI) from the cell occurs through a transmembrane protein encoded by gene *chrA* in *P. aeruginosa*.

Lab-scale bioreactors were employed to study chromium biosorption by *P. aeruginosa* as far back as 2002. Ganguli and Tripathi (2002) evaluated the Cr(VI) reducing ability of *P. aeruginosa* A2Chr in two types of laboratory scale bioreactors compared with batch culture. In the first bioreactor, *P. aeruginosa* A2Chr was entrapped in a dialysis sac, immersed in effluent comprised of varying concentrations of Cr(VI); in the second bioreactor, the *P. aeruginosa* A2Chr was entrapped in an agarose-alginate biofilm, mounted onto a rotating biological contactor. Lower reduction rates of chromate were observed in the dialysis sac and this was attributed to the effluent, which may have clogged the pores and complicated free diffusion across the membrane. It was also observed that the artificial biofilm mounted on the rotating biological contactor was more adept at reducing Cr(VI) up to high concentrations. It appeared that the alginate-agarose film provided some protection for the bacteria against the toxicity of Cr(VI), as had been previously suggested by Xu et al. (1996). Meanwhile, Kang et al. (2007) investigated the biosorption of chromium ions Cr (III) and Cr (IV) on the cell surface of *P. aeruginosa* and found the chromium binding sites on the bacterial cell surface were most likely carboxyl and amine groups. In the case of Cr(VI) biosorption by *P. aeruginosa*, the reduction and adsorption of Cr(VI) occurred coincidently in an abiotic process. This finding is crucial because most bacteria found in the subsurface exist in nutrient-deficient conditions under natural conditions (Fein et al., 2002). The biosorption of Cr(VI) by *P. aeruginosa* occurs through two processes: the first being the reduction of Cr(VI) to Cr(III) by reductive functional groups and second being chromium ion removal from wastewater by adsorptive functional groups of *P. aeruginosa*. The electrostatic interaction between the charged surfaces of bacteria and chromium ions provides the adsorptive property needed to remove the chromium ions. Results from Kang et al. (2007) indicate that bacterial functional groups of *P. aeruginosa* can act as reductive and adsorptive sites in biosorption of heavy metals.

Xu et al. (2009) evaluated the Cr(VI) resistance and reduction by *P. aeruginosa* and found that the bacteria was able to reduce the metal ion despite toxic effects on the cell. It was also noted that the presence of Zn^{2+} inhibits the Cr(VI) reduction while Cu^{2+} significantly stimulates the Cr(VI) reduction. It was speculated that for the stimulatory mechanism of Cr(VI) reduction by Cu^{2+}, Cu^{2+} may in some cases, act as a shuttle for electrons between protein subunits. The presence of SO_4^{2-} and NO_3^- had no effect on the Cr(VI) reduction. This was further supported by the fact that aerobic chromate reduction was generally not affected by sulfate and nitrate (Wang and Xiao 1995; Garbisu et al. 1998; Liu et al. 2006). Anaerobic chromate reduction, however is affected by nitrate and sulfate ions due to the fact that they may compete with chromate as anaerobic electron acceptors. Since the reduction was observed over a range of redox potential (from +3 mV to +91 mV), it was also deduced that a low redox potential was not necessary for Cr(VI) reduction by *P. aeruginosa.* Previous work involving Cr(VI) with other bacteria such as *Agrobacterium radiobacter* (Llovera et al., 1993) and *E. coli* (Gvozdyak et al., 1986) had indicated that a low redox potential was required for Cr(VI) reduction. Bakiyaraj et al. (2014) isolated *Bacillus sp.* and *Pseudomonas sp.* from chromium contaminated soil and found a significant difference in the growth rates of the isolates at different concentrations of chromium. They found that as the concentration of chromium increased, the growth of the bacteria decreased. This was attributed to the fact that when initial concentrations of chromium were high, less biomass was available to remove the chromium from solution, since the volume of biomass was kept constant. The results coincide with those of Raghuraman et al. (2013) and Basu et al. (2014) who observed the inverse relationship of chromium removal and initial concentrations of chromium. Singh et al. (2013) examined the effectiveness of Cr(VI) and cadmium removal in a small scale reactor using *P. aeruginosa*. They found that *P. aeruginosa* was able to effectively remove the metal contaminants and also found a synergistic effect of the bacteria with Fe^0. Ojiagu et al. (2018) more recently evaluated the biosorption potential of *P. aeruginosa* strain ANSC (non-genetically modified) for Cr(VI) from solution and found the adsorption behavior was highly dependent on temperature and pH with optimum pH at 2 and optimum temperature at 60°C. They also found that the rate of metal adsorption increased with the use of surfactants (such as Triton X100, Tween 80 and toluene) to permeabilise the outer and/or cytoplasmic membrane of the bacteria.

Cadmium

Cadmium is one of the most dangerous metal ions discharged in industrial effluents. Due to its corrosion-resistant properties, it is used mainly to plate metal parts in electronics, automobiles, marine and aerospace industries (Herrero et al., 2005; Kermani et al., 2010). It is also used in PVC production, nickel cadmium battery manufacturing, printing and dyeing and nuclear power plants. The metal is characterised by high stability and toxicity. It does not degrade in nature and hence, once released to the environment, will remain in circulation. A major proportion of cadmium release to the environment is caused by dumping and incinerating cadmium-polluted waste (Tang et al., 2018). The situation is further aggravated by the use of phosphate fertilizers in agricultural activities resulting in severe cadmium pollution in the environment. Cui et al. (2017) reported over 10% of rice grain samples across China currently exceed the allowable cadmium limit of 0.2 mg/kg. Cadmium can bind with essential respiratory enzymes in humans causing oxidative stress and in more severe cases, cancer (Nies, 2003; Banjerdkij et al., 2005).

In recent years, efforts have been underway to isolate heavy metal-resistant bacterial strains to treat cadmium. Early research has found cadmium to be excluded from the cell surface of *Staphylococcus aureus* and *Escherichia coli* (Novick and Roth, 1968; Mitra et al., 1975). Hassen et al. (1998) found that *P. aeruginosa* can detoxify Cd^{2+} through the production of intracellular cadmium-binding proteins. Chovanová et al.

(2004) evaluated bacterial colonies isolated from sewage sludge contaminated by cadmium ions. In the bacterial community, it was found that short cadmium resistant gram-negative rods were predominant. Analysis identified the eight isolates to six species: *Alcaligenes xylosoxidans*, *Comamonas testosteroni*, *Klebsiella planticola*, *Pseudomonas putida*, *Pseudomonas fluorescens*, and *Serratia liquefaciens*. A correlation was found between the cadmium removal efficiency and the amount of additionally synthesized proteins in the cell fractions suggesting a direct relationship between the two. Kermani et al. (2010) showed that *P. aeruginosa* isolate could efficiently remove a significant amount of cadmium in solution (94.7% in 30 mg/L of Cd^{2+} solution within 60 min). The bacteria were isolated from the active sludge of a food processing plant and was exposed to mutagenic agents of acridine orange and acriflavine. The mutated bacteria were shown to perform significantly better than the wild type. The results were consistent with previous work on *P. aeruginosa* where both living and non-living cells of the strain isolated by Zeng et al. (2009) was observed to remove Cd^{2+} from solution. Also expounded in the work was the observation that the living cells had higher Cd^{2+} removal rates than non-living cells.

Perhaps one of the more interesting works on biosorption of cadmium by *P. aeruginosa* can be found in Izrael-Živković et al. (2018). They reported *P. aeruginosa* san ai (environmental isolate from mineral cutting oil) as promising candidate for treating up to 7.2 mM of cadmium. Using results from shotgun proteomics approach, they speculated that *P. aeruginosa* san ai develops a complex mechanism to adapt to cadmium. The mechanism encompasses a complex combination of extracellular biosorption, bioaccumulation, biofilm formation, controlled production of siderophore, enhanced respiration and modified protein profile. The isolate also evolved several defense strategies at the intracellular and extracellular levels: first extracellular binding of cadmium to exopolysaccharide released in culture broth, followed by binding of cadmium to cell surface, increased production of metalloproteins, especially denitrification proteins and finally metabolic adaptation to increased concentration of cadmium within the cell. Tang et al. (2018) used CadR (a cadmium-specific binding protein) on the surface of *P. aeruginosa* with chromosomal expression and found that the genetically engineered (GE) *P. aeruginosa* exhibited extraordinary genetic stability and high adsorption capacity of Cd(II). Their results show that CadR anchoring on the outer membrane by ice nucleation protein (INP) significantly enhanced Cd(II) adsorption on the cell.

Lead

Lead is a mutagenic metal causing severe effects on human beings such as neurodegenerative impairment, renal failure, reproductive damage and cancer (Fowler, 1998; Tong et al., 2000; Watt et al., 2000; Lam et al., 2007; Shahid et al., 2012; Naik and Dubey, 2013). Lead content in the environment has increased over 1000-fold over the past three centuries as a consequence of anthropogenic activities (Naik and Dubey 2013). Lead can be found predominantly in wastes from mining, sewage sludge, metal refining, power plant and incineration operation. Despite the toxic nature of lead, some bacteria employ a variety of protective mechanisms to survive high levels of lead without any impact on their growth. The strategies of these bacteria to resist high concentrations of lead include efflux mechanisms, extracellular sequestration, biosorption, precipitation, alteration in cell morphology, enhanced siderophobe production and intracellular bioaccumulation (Naik and Dubey, 2013). The mechanisms in which *P. aeruginosa* have been reported to exhibit are: intracellular bioaccumulation, extracellular sequestration, surface biosorption and alteration in cell morphology. In terms of intracellular bioaccumulation, *P. aeruginosa* (BmtA) has been reported to encode metallothioneins for maintaining cytosolic metal homeostasis (Turner et al., 1996; Blindauer et al., 2002; Liu et al., 2003) Metallothioneins are specific metal binding proteins which facilitates the bioaccumulation of toxic metals inside the cell (Naik and Dubey, 2013). In addition, *P. aeruginosa* strain WI-1 isolated from Mandovi estuary was found to contain bacterial metallothionein (BmtA) to alleviate lead toxicity (Naik et al., 2012). The strain was also shown to resist 0.6 mM of lead nitrate by bioaccumulating 26.5 mg lead/gram dry weight cells intracellularly. As for the surface biosorption mechanism, *P. aeruginosa* PU21 was shown to adsorp lead, copper and cadmium from aqueous metal solution (Chang et al., 1997). This was due to the numerous negatively charged chemical groups present on the bacterial cell surface, and since *P. aeruginosa* is a Gram-negative bacterium, the phosphate group would typically serve as the main metal binding site at the cell wall (Gadd and White, 1993).

Gabr et al. (2008) studied the optimum conditions for biosorption and bioaccumulation of lead and nickel using living and non-living *P. aeruginosa* ASU 6a isolated from El-Malah canal in Assiut, Egypt. It was found that the binding capacity of the living cells were significantly lower than that of the non-living cells. The mechanism of biosorption was confirmed (using IR and pH experiments) to be mainly ionic interactions and complex formation between metal ions and acidic sites in the cell wall of the bacteria. Naik and Dubey (2011) reported that a lead-resistant *P. aeruginosa* strain 4EA from soil contaminated with car battery waste from Goa, India resisted 0.8 mM of lead nitrate by biosorption of 11% wt of lead on the cell surface. The strain also exhibited evidences of alteration in cell morphology where the size of the cell reduced in response against the toxic lead. Konig-Péter et al. (2014) studied the potential of lyophilized *P. aeruginosa* PAO1 to adsorb Cd(II) and Pb(II) ions from aqueous solution. The optimum pH values for Cd(II) was found to be between 5.0 – 6.0 and 5.0 for Pb(II). Bojórquez et al. (2016) studied the biosorption of cadmium and lead with *P. aeruginosa* and *Enterobacter cloacae* and found that both strains showed high biosorption rates for lead but low rates for cadmium. *P. aeruginosa* was also notably more efficient in the sorption of lead than *E. cloacae* (82.5% vs 72.2%). The mechanism at work was speculated to be adsorption to the bacteria cell wall followed by active transmembrane transport, chelation/sequestration of metal ions and intracellular accumulation (King et al., 2007). More recently, Karimpour et al. (2018) examined the adsorption efficiency of cadmium and lead in aqueous solutions using live and dead biomass of *P. aeruginosa* and reported high adsorption rates for both metals. The highest adsorption efficiency for cadmium and lead were reported to be 87% and 98.5% respectively, (for a dead cell mass of 125mg, pH 7, temperature 35 °C and contact time 90 minutes). Freeze dried bacterial isolate *P. aeruginosa* 2103 from sewage water was found to be highly resistant to Pb^{2+} ions (Rasmey et al., 2018). The best conditions were found to be at equilibrium pH of 6 and a contact time of 30 min at room temperature.

Other Heavy Metals

Besides Cr, Cd and Pb, various studies have reported on the ability of *P. aeruginosa* in biosorption of other heavy metal elements such as Al, Co, Cu, Hg, Mn, Ni and Zn (refer to Table 4.1). The maximum adsorption capacity of *P. aeruginosa* biomass/bioproduct range from 5.25 to 400 mg of metal ions per g of dry weight with optimum pH of 5.0 to 9.0. Biosorption of these metal ions is dependent on pH of the aqueous medium due to the presence of functional groups in the cell wall and the chemical configuration of the metal. By increasing the pH, the cell surface becomes negatively charged, causing deprotonation of the amines, amides and carboxylic acid functional groups. This situation favours electrochemical attraction and sorption of cations. Increase in metal uptake in higher pH may also resulted from a more efficient competition between cations and H^+ for binding sites on the bacteria (Silva et al., 2009). The presence of other cations such as Na, K or Ca was also shown to increase sorption of Hg, Ni or Cu, in contrast to Cd and Pb (Chang and Hong, 1994; Sar et al., 1999). In the case of Hg, the presence of Na was thought to provide accessible sites on mercury binding proteins through a favourable conformational change, enabling high mercury capacity and sorption. It was then suggested that biosorption of mercury is not merely a simple ion exchange process, but an affinity-based selective adsorption (Chang and Hong, 1994). Moreover, with their elevated polarizing power, Ni and Cu can displace Na, K and Ca that bind more favourably to cellular ligands with N or O donor atoms. This increases available binding sites for both Ni and Cu, and thus enhancing metals uptake by the bacteria. (Sar et al., 1999).

Table 4.1. *Pseudomonas aeruginosa* biosorption of other heavy metal elements

Ele-ments	*P. aeruginosa* biomass/bioproduct	Experimental Medium Conditions	Maximum adsorption capacity (mg/g dry weight)	References
Al^{3+}	Heat-dried, grounded *P. aeruginosa* A22	3.0 mM, pH 5.0	60.7	Texier et al., 1999
	Heat-dried, immobilized on Chromosorb 106	20 ppm, pH 6.0	10.5	Tuzen and Soylak, 2008
Co^{2+}	Heat-dried, immobilized on multiwalled carbon nanotubes	1.0 mg metal ions in 50 mL, pH 9.0	6.06	Tuzen et al., 2008
Cu^{2+}	Lyophilized *P. aeruginosa*	4.0 mM, pH 7.0	137.6	Sar et al., 1999
	Live *P. aeruginosa* AT18	49 mg L^{-1}, pH 6.25	113.64	Silva et al., 2009
	Live *P. aeruginosa* MCCB 102	1.0 mM	70.4	Zolgharnein et al., 2010
Hg^{2+}	Inactivated steam-sterilized *P. aeruginosa* PU21	0.05 – 2.5 mM, pH 7.4	400	Chang et al., 1994
	Carboxylesterase E2 *P. aeruginosa* PA1	1.0 mM, pH 7.4	122.4	Yin et al., 2016
Mn^{2+}	Heat-dried, immobilized on multiwalled carbon nanotubes	1.0 mg metal ions in 50 mL, pH 9.0	5.83	Tuzen et al., 2008
	Live *P. aeruginosa* AT18	60 mg L^{-1}, pH 7.0	22.39	Silva et al., 2009
Ni^{2+}	Lyophilized *P. aeruginosa*	4.0 mM, pH 8.0	265	Sar et al., 1999
	Heat-dried, immobilized on multiwalled carbon nanotubes	1.0 mg metal ions in 50 mL, pH 9.0	5.25	Tuzen et al., 2008
	Live *P. aeruginosa* ASU 6a	0 - 160 ppm, pH 7.0	70	Gabr et al., 2008
	Heat-dried *P. aeruginosa* ASU 6a	0 - 160 ppm, pH 7.0	113.6	Gabr et al., 2008
Zn^{2+}	Live *P. aeruginosa* AT18	70 mg L^{-1}, pH 7.0	77.5	Silva et al., 2009
	Live *P. aeruginosa* MCCB 102	1.0 mM	65.9	Zolgharnein et al., 2010
	Lyophilized *P. aeruginosa* ASU 6a	50 mg L^{-1}, pH 6.0	83.3	Joo et al., 2010

Mathiyazhagan and Natarajan (2011) reported that bacteria *P. aeruginosa* and *T. ferrooxidans* isolated from the waste dump of magnesite and bauxite mines showed high resistance to heavy metals (Zn, Mn, Cu, Cr and Hg). The results of the work revealed that *T. ferrooxidans* reduced and/or absorbed some of the heavy metals from the mines such as Cd, Ca, Zn, Cr, Mn, and Pb while *P. aeruginosa* absorbed most of the metals that *T. ferrooxidans* could not absorb. The work indicated that in certain conditions, *P. aeruginosa* was a better agent for biosorption than *T. ferrooxidans*. *T. ferrooxidans* is a well-known species for treating heavy metal contaminated tailing and soils, and is also used in industrial bio-leaching of metal sulfide and uraninite ores (Kelly, 1988; Straube et al., 2003). *P. aeruginosa* was employed in the effluent of steel, pulp and paper and leather industries respectively and found to yield acceptable results for Cr and Fe (Awasthi et al., 2015). *P. aeruginosa* isolate from petroleum refinery effluent was also found to yield higher values of metal reduction compared to other pure isolates such as *Staphylococcus aureus*, *E. coli*, *Proteus vulgaris* and *Klebsiella pneumonia* (Oaikhena et al., 2016). The pure isolates were inoculated into varying concentrations of Cd, Cr, Ni and Zn.

It is also interesting to note that metals uptake using heat-dried or lyophilized cells is more efficient than living *P. aeruginosa* cells. Sorption by heat-dried or lyophilized cells involved passive uptake that is generally energy independent, whereas in active uptake by living cells, energy is required for metal transfer across the cell wall into the cytoplasm and the presence of H^+ that competes for binding sites (Gabr et al., 2008). By pre-treating the cells with heat, the bacterial cell wall will be broken, and this exposes latent binding sites, thus increases overall metal uptake (Galun et al., 1987; Srinath et al., 2002). Similarly, as observed for *P. aeruginosa* biomass, treatments such as heat-drying or freeze drying could contribute to an increase in both surface area and exposure of intracellular binding sites in the cells. These factors taken together with the absence of competition by H^+ ions constantly produced by viable cells, attributed to the improvement in adsorption capacity seen in pre-treated cells (Gabr et al., 2008).

Nevertheless, some studies reported the negative effect of heat pre-treatment in biosorption of metals. It was shown that thermal treatment altered the physical and chemical properties of the cells by destroying some surface-active components, leading to reduced intracellular uptake capability (Park et al., 2004; Chen et al., 2005).

Radionuclides

Radionuclides are species of unstable atoms that release radiation as they undergo radioactive decay and may occur in nature or are artificially produced. Naturally-occurring radionuclides are categorized into three groups namely: primordial radionuclides, secondary radionuclides and cosmogenic radionuclides (Eisenbud and Gesell, 1997). Primordial radionuclides exist since the formation of earth and have yet to decay completely due to their long half-lives in the order of billions of years. Secondary radionuclides have shorter half-lives than primordial radionuclides and are derived from the decay chain of primordial isotopes ^{238}U, ^{232}Th and ^{235}U. Naturally-occurring radionuclides are also generated by cosmic radiation which interacts with the earth's upper atmosphere to produce radioactive nuclides. Majority of cosmogenic radionuclides have very short lives except ^{14}C and ^{3}H with half-lives of 5730 years and 12.3 years respectively. The concentrations of these naturally-occurring radionuclides can be amplified through human activities and industrial processes such as mineral mining, gas and oil production, coal combustion, water treatment and purification, and paper and pulp manufacturing, which could rise to a dangerous level that may jeopardize both human and the environment (Ojovan and Lee, 2014). Radionuclides can also be synthesized artificially using nuclear reactors, radionuclide generators or particle accelerators, and are released into the environment through nuclear weapon testing, nuclear energy activities and accidents at nuclear power plants (Das, 2012).

The presence of radionuclides has been detected in soils, sediments and water around the world, mainly due to anthropogenic activities. These radionuclides can be present in many forms, often determined by the physiognomies of the surrounding ecosystem. Some radionuclides are associated with colloids and these are usually mobile, while others can form complexes with soil organic and inorganic materials or precipitate as pure or mixed solids, thus being retained in sinks such as soils and sediments (Salbu et al., 2015). The immobility of radionuclide complexes and migration of colloids cause accumulation on the upper layer and subsurface of soil, and these radionuclides can be easily integrated into the food chain, causing threat to human and animal health (Das, 2012). Most radionuclides are geochemically stable and insoluble in lower oxidation states, however in nature, they tend to exist in higher oxidation states such as uranium(VI), technetium(VII), technetium (IV), thorium(IV) and plutonium(V). These radionuclide species are contaminants in surface water and groundwater as they are highly soluble, leading to mobilization of toxic compounds into drinking water and crop irrigation (Munter, 2013). The release and persistence of radionuclides in the environment can lead to a long-term external irradiation or internal contamination via the food chain or drinking water among the populations living in these contaminated areas. Long-term exposure to radionuclides either through radiation, ingestion or inhalation, has been known to cause multiple health effects such as increased risk of lung, bone and soft tissue cancer, leukaemia, kidney failure, affecting fertility and development of foetus, and genetic damage (Leiterer et al., 2010; Raabe, 2010; Yadav et al., 2017).

The interaction of *P. aeruginosa* with uranium will be the focus in this subsection. In a study by Strandberg et al. (1981), rapid uptake of uranium from solution by *P. aeruginosa* and *Saccharomyces cerevisiae* cells were reported. Uranium accumulation by *P. aeruginosa* was extremely rapid (<10s), occurred intracellularly and was not affected by environmental parameters. In contrast, uranium accumulation by *S. cerevisiae* was slightly slower and can be enhanced by chemical pre-treatment of the cells. Despite metal uptake by either organism was not dependent on metabolism, the mechanism of uranium uptake differs between *P. aeruginosa* and *S. cerevisiae*. Consistent with metal biosorption occurs through complexation of positively charged metal ions with negatively charged reactive sites on the cell surface, uranium was seen accumulated as needle-like fibrils in a layer on the surfaces of *S. cerevisiae* cells, with little or no uranium found inside these cells. Conversely, dense intracellular uranium deposits were found in *P. aeruginosa*, suggesting possible uranium localization inside the cells by heavy metal-binding protein metallothionein, though this has yet to be validated. More recently, it was shown that sequestration of uranium as a possible cellular mechanism for uranium tolerance in various bacteria, and this process resulted in intracellular accumulation of uranium precipitates possibly to limit uranium toxicity (Suzuki and Banfield, 2004; Martinez et al., 2007). Consistent with these, a uranium-resistant *P. aeruginosa* strain isolated from uranium mine waste was capable of accumulating large amount of uranium (maximum of 275 mg U/g cell dry weight) and precipitate it within the cell envelope region. Precipitation of accumulated uranium as uranium phosphate within the cell envelope is believed to be vital in protecting biologically important molecules in the cytoplasm (Choudhary and Sar, 2011). The rapid uptake of uranium could be attributed to the strong affinity of uranium towards anionic sites on the surface of the cells and is divided into two distinct adsorption phases (Strandberg et al., 1981; Philip et al., 2000) The initial phase is rapid and involves ion-exchange processes and biosorptions of metal ions onto the cell wall. The second phase comprises a gradual uptake of uranium, which is a slow process and may be due to biologically mediated transfer of metal ions into the cytoplasm (Philip et al., 2000).

Besides living cells, several works have described the use of *P. aeruginosa* biomass or bioproduct in biosorption of uranium. Hu et al. (1996) reported that heat-killed, permeabilised and unreconstituted lyophilized cells of *P. aeruginosa* strain CSU were capable of binding uranium, comparable to living cells. *P. aeruginosa* CSU biomass was significantly more sorptive towards uranium (approximately 100 mg U/g dry biomass) than several algal or fungal biosorbents and its rate of adsorption is dependent on pH and the presence of organic solvents. In Kazy et al. (2008), the ability of *P. aeruginosa* BU2 extracellular polysaccharide (EPS) to remove uranium from aqueous solution was reported. The EPS exhibited fast saturating uranium sorption which can be enhanced by increasing solution pH and was unaffected by interfering cations and anions, except iron and thorium. Additionally, *P. aeruginosa* were able to produce siderophores that are capable to change pH and enhance the chelation of uranium and thorium (Bouby et al., 1998a, 1998b) Recently, a siderophore from *P. aeruginosa* was shown to complex with 68% of uranium and 65% of thorium from residual Th-U concentrate, suggesting its application as potential ligand in recovery of these metals from the environment (Desouky et al., 2016).

Rare Earth Elements

Rare earth elements (REEs) are a set of metals comprised of fourteen lanthanide elements, yttrium and scandium, with unique chemical and physical properties that make them important for various established and emerging technologies. Among the uses of REEs are as catalysts for petroleum refining and automotive, strong magnets for electrochemical devices, additives in glass, lenses and ceramic production, alloys in fuel cells, NiMH batteries and metallurgy sector, phosphors for display screens, fluorescent lamps and medical imaging, lasers, fibre optics, as well as in water treatment, as fertilizer and tracers in medical application (Van Gosen et al., 2014; Rim, 2016). Despites their abundance in nature, REEs are typically dispersed and not often found as concentrated ore bodies due to their geochemical properties. Moreover, they are mostly found together in nature or associated with uranium/thorium mineralization, making them difficult to be separated from one another (Haque et al., 2014). As the demands and utilization of REEs in industries continue to grow, more REEs are being introduced into the environment through activities such as mining and processing. Mining and processing of REEs often produced significant amounts of waste and pollution-intensive materials such as tailings, which typically contain potentially harmful contaminants that pose a significant environment risks affecting water discharges, tailing management and air emissions (Haque et al., 2014; Huang et al., 2016).

Additionally, exposure to higher concentration of REEs was known to cause cytogenetic effects, organ-specific toxicity and inhibition of growth (Pagano et al., 2015; Rim, 2016).

Removal and/or recycling of REEs from process streams using conventional methods have been proven to be either inefficient or expensive for applications on large volume dilute wastewater. To overcome these challenges, attention has been focused on the capabilities of microorganisms and biopolymers to bind to metals. Microbial biosorption is very efficient in removal of REEs from very dilute solutions and in some cases, are highly selective for a certain metal of interest (Texier et al., 1999; Das and Das, 2013). In an earlier study, significant differences on biosorption capabilities were observed among bacteria when 33, 70, 114 and 144 μmol/g of La^{3+} were removed by *B. cereus*, *E. coli*, *B. subtilis* and *P. aeruginosa*, respectively, demonstrating the superiority of *P. aeruginosa* in biosorption of lanthanum. Additionally, electron microscopy images also showed precipitation of needle-like crystalline on the bacterial surface, with no evidence of uptake into the cytoplasm (Mullen et al., 1989). It was later shown that adsorption of lanthanum on the surface of *P. aeruginosa* was accompanied by partial release of calcium and magnesium ions from the cell wall, consistent with ion exchange mechanism observed in the biosorption of uranium (Philip et al., 2000).

Further study by Texier et al. (2000) has shown that binding of metal ions happens on acidic moieties. Using potentiometric titration, they discovered two distinct waves indicating the presence of at least two binding sites on the cell wall of *P. aeruginosa*, corresponding to strong and weaker acidic groups with pKa of 2.8 ± 0.2 and 6.1 ± 0.2, respectively. Treatment with base solutions neutralized all the acidic groups in the biomass, suggesting that the cell wall of *P. aeruginosa* would contain at least two types of strong acidic groups, potentially phosphate and carboxyl moieties, and two types of weaker acidic groups that contribute to the high adsorption capacities demonstrated by this biomass. Interestingly, it was also shown that *P. aeruginosa* is capable of selective biosorption of lanthanide ions (Texier et al., 1999). The lanthanide biosorption equilibrium followed the Brunauer-Emmett-Teller isotherm model, indicative of a multilayer adsorption. In single cation solutions, the maximum adsorption capacities were 397 μmol/g for lathanum, 326 μmol/g for ytterbium and 290 μmol/g for europium. This showed that the bacterial cell wall provides preferential sites for La which are not accessible for Eu and Yb. It was speculated that the cavity diameter at these sites would be too large to bind Eu and Yb with a corresponding degree of stability. However, in mixed-cation solutions, the biosorption of La was strongly affected by the presence of Yb (-77%) and Eu (-71%). The biosorption of Yb was slightly affected by the presence of La (-42%), whereas uptake of Eu was unaffected by La. It was also observed that the uptake of Eu decreased by half in the presence of Yb and vice versa, showing a preferential biosorption of Eu^{3+} = Yb^{3+} > La^{3+}. Similarly, preferential biosorption of lanthanide ions was also reported for immobilised *P. aeruginosa* cells, proposing possible application of this biomass in removal and separation of lanthanide ions from aqueous effluents (Texier et al., 2002). In addition, it was found that the presence of Al^{3+} inhibited the uptake of lanthanide ions especially La^{3+} and Yb^{3+} by *P. aeruginosa*, whereas no significant effects were detected in the presence of anions. Therefore, considerations to include preliminary treatment to remove aluminium ions may be necessary in the design of wastewater treatment using biosorption (Texier et al., 1999).

Precious Metals

Large-scale extraction and refining operations of precious metals has resulted in the production of vast volume of wastewaters, which normally contain substantial amounts of valuable metal (Mack et al., 2007). Due to increasing demands globally, recovery and recycling of these precious metals from wastewater has been a subject of wide interest. One of the most promising methods in recovery of these valuable metals is biosorption. However, the biosorption of precious metals is usually very different to that of base metals. For example, in platinum group metal (PGM) refinery wastewaters which typically consist of high chloride concentrations and low pH, PGMs are present in the form of anionic chloro-complexes. Consequently, this may affect the adsorption mechanism involved and the binding of metal species on the biomass. Biosorption for precious metals occur mostly through chelating instead of ion exchange and at low pH (1.0-3.0), whereas for base metals, adsorption occur at higher pH of 3.0 to 7.0 (Sheng et al., 2004; Godlewska-Żyłkiewicz and Kozłowska, 2005). Various biosorbents had been extensively described for the adsorption of precious metals including chitosan, tannins, algae, fungi, yeasts and bacteria (Mack et al., 2007; Won et al., 2014). However, information on the ability in biosorption of precious metals by *P. aeruginosa* had only been focused on gold and silver.

In a study by Tsuruta (2004), various species and strains of bacteria, actinomycetes, fungi and yeasts were screened for their gold adsorbing ability from a solution containing hydrogen tetrachloroaurate (III) or sodium dicyanoaurate (I). These solutions were commonly used in medical and ceramic materials and for plating, respectively and present in different oxidation stage. From this screening, it was found that most of the Gram-positive bacteria, actinomycetes, fungi and yeasts displayed lower gold biosorption ability than the Gram-negative bacteria. Of the Gram-negative bacteria, high gold biosorption from hydrogen tetrachloroaurate (III) were observed in *Erwinia herbicola* (360±1 μmol/g dry weight cells), *P. aeruginosa* (340±5 μmol/g), *P. maltophilia* (337±0 μmol/g) and *Acinetobacter calcoaceticus* (336±1 μmol/g). Biosorption of gold from sodium dicyanoaurate solution however, is far lower in all the microorganisms tested. The superior ability of gold biosorption observed in Gram-negative bacteria is thought to be contributed by the positive charge of its cell surface. As gold(III) exists as a negatively charged complex ion in solution, it can bind more strongly to the positively charged cell surface of Gram-negative bacteria as compared to negatively charged Gram-positive bacterial cell surface. Using *P. maltophila*, Tsuruta (2004) also reported that gold biosorption occurred rapidly and was affected by the pH of the solution, cell amounts and external gold concentration. Additionally, a study by Husseiny et al. (2007) has shown the ability of *P. aeruginosa* in extra-cellular biosynthesis of gold nanoparticles (Au NPs). Bacterial cell supernatant of *P. aeruginosa* ATCC 90271 and two clinical isolates of *P. aeruginosa* were able to reduce gold ion into well-dispersed Au NPs in the range of 15-30 nm, suggesting the possible application of extra-cellular reduction for the development of an easy bioprocess for synthesis of Au NPs.

In Mullen et al. (1989), a study on the biosorption of Ag in four bacteria, namely *B. cereus*, *B. subtilis*, *E. coli* and *P. aeruginosa* were reported. It was shown that bacterial removal of Ag^+ from the solution was very efficient ranging from 89 to >99% of the total Ag^+ at initial concentrations of 1 to 0.01 mM. Removal of silver from the solution was independent of the bacterial species, though *P. aeruginosa* displayed lower biosorption of Ag than the other bacterial. Moreover, electron micrographs of silver-treated cells shown that Ag was concentrated in discrete particles in the cell walls, and in about 1% of the cells, were found in the cytoplasm adjacent to the plasma membrane. Besides the use of whole cells, a copper protein involved in the *P. aeruginosa* electron transfer chain, namely azurin, was reported to preferentially bind to Ag^+ in its reduced state (Tordi et al., 1990). In reduced azurin, Ag^+ completely displaces the Cu^{2+} from its native binding site and this effect is very specific for silver ions. Treatment with excessive Ag^+ indicated a second binding site for Ag^+ on the protein which can bind to Cu^{2+} and Hg^{2+} with similar affinities. A kinetic study on this interaction further showed that the binding of Ag^+ to reduced azurin is four times faster than binding to the apo-form, suggesting that conformational rearrangement of the active site of the protein is necessary for metal binding (Naro et al., 2000).

SCALING-UP: CHALLENGES AND LIMITATIONS

In recent years, biosorption has been increasingly touted as an alternative to conventional metals removal processes involving low operating costs, reasonable efficiency and low sludge volume. Despite progress made over decades of research work, most biosorption processes are still at the laboratory scale (Fomina and Gadd, 2014). Pilot scale studies are essential for the application of the process at the industrial scale. Through pilot studies, at a higher scale-up factor from the laboratory scale, costs can be more accurately estimated, and the tests can more closely depict the process at full industrial scale.

The choice of bioreactors in scaling-up depends largely on factors such as hydrodynamic conditions, bacteria growth conditions and rate of mass transfer. Typical types of reactors that can be used are fixed-bed reactors (FXR), stirred tank reactors (STR), fluidized-bed reactors (FBR), and AirLift reactors (ALR). Of these reactors, fixed bed reactors are known to be effective, economical, convenient and easy to scale test beds. Other reactors, such as stirred tank and fluidized-bed reactors are used to a lesser extent for biosorption (Das and Das, 2013). Stirred tank reactors are useful when biosorbent exists in the powder form and they usually involve higher capital and operating costs. Fluidized-bed reactors, on the other hand, use high flow rates to maintain the biosorbent particles in suspension. *P. aeruginosa* has been reportedly tested using fixed-bed reactors on a laboratory scale. Ajao et al. (2011) a mixed culture of *P. aeruginosa* and *Bacillus subtilis* to remove toxic components from industrial textile effluents which includes Cr(VI). Fixed-bed reactor was also employed in the work of Ganguli and Tripathi (2002) who evaluated the Cr(VI) reducing ability of *P. aeruginosa* A2Chr in two types of laboratory scale bioreactors compared with batch culture.

Throughout the years, a number of proposed processes involving biosorption have been patented for commercial application. Pilot installations and several industrial scaled units have been reported (Tsezos, 1999). A few biosorbents that were developed as commercial products for removing metals from solution include: AlgaSORB™ (made from freshwater microalga), B. V. Sorbex Biosorbent (made from macroalgae), AMT-Bioclaim™ (manufactured using immobilised *Bacillus* sp.), Bio-Fix Biosorbent (made from algae immobilised in porous polypropylene beads) and Rahco Bio-Beads (uses peat moss immobilised within an organic polymer) (Vijayaraghavan and Yun, 2008; Park et al., 2010; Fomina and Gadd, 2014). However, the lack of success of these products begs for critical evaluation. Firstly, the existing technologies are already in place treating large volumes of effluents to date. The prevailing conception is one that has embraced these working solutions albeit known to be less sustainable. Take ion-exchange for instance, where its commercial success is largely due to its proven results at the industrial scale. The biosorption process however, still has its image plagued by the complexity of functional groups involved in the process and the availability of biosorbents of choice. Another reason is the low technology readiness level of biosorption processes. For instance, the lack of understanding of mechanisms, kinetics and thermodynamics involved, impeded the objective evaluation of AlgaSORB™ and AMT-Bioclaim™ commercialised in the early 1990s as a viable process (Kotrba, 2011; Fomina and Gadd, 2014). The lack of selectivity and lower robustness of biosorption processes compared to conventional methods have also been cited as main reasons limiting the commercialisation of biosorption (Eccles, 1999). Gadd (2009) deemed suspended biomass as ineffective and not durable for repeated long-term use. There also lies the conundrum where any attempts to manipulate or treat biosorbents tend to offset its advantages in the first place. For instance, chemical or biological modification to improve properties, alternately increases the cost of biosorbents and generates toxic wastes that may also be biohazardous.

Furthermore, complications arise when considering the implementation of biosorption at full industrial scale. Biosorption is strongly dependent on various parameters (pH, temperature, ionic concentration etc.), most of which has been explained in previous sections. The difficulty in applying biosorption on real wastewaters lies in the huge volume and complexity of the wastewater matrix. Hence, at full scale, it would be difficult to control the process to suit specific biosorbents. Many bacteria, including *P. aeruginosa* require an optimum pH range to achieve maximum uptake. Considering the immense volume of wastewater involved, it is rather impractical to alter the pH of the solution. Large amounts of pH modifiers would have to be added to control the pH of the solution, implying costs that may render the entire process unfeasible. Also, studies have shown that the presence of certain metal ions may negatively affect the biosorption of other competing metal ions. Therefore, biosorption may only be suited when removal of metal ions is non-specific. Hence, biosorption may be more suited as a process to improve the overall quality of wastewater rather than removal of a specific metal ion.

It may prove to be worthwhile to start applying biosorption to treat laboratory waste effluent. Laboratories, unlike industrial plants often store highly complex effluent of unknown chemical composition due to the various tests conducted over extended periods of time. Wastewaters coming from chemical laboratories typically consist of metal ions in lower milligram levels or less but are above the permissible limits to be discarded into sewers. In efforts to scale up the biosorption process from laboratory scale, laboratory waste effluents may serve as a workable test bed, to assess the feasibility of applying the process to larger volumes of wastewaters and the intricacies resembling that of actual wastewater matrix. Laboratory waste effluents also may be suitable since no specific metal ion is targeted and hence, the objective is the overall improvement of the wastewater content i.e., reduced overall heavy metals concentrations. Hybrid *Sargassum*-sand sorbent have been reported to successfully clean ICP-OES (Inductively coupled plasma-optical emission spectrometer) wastewater consisting of 14 metal ions (Vijayaraghavan and Joshi, 2013).

FUTURE PERSPECTIVES

Tuzen et al. (2008) investigated the use of *P. aeruginosa* immobilised multiwalled carbon nanotubes for biosorption of metals. Carbon nanotubes are one of the building blocks of nanotechnology, with superior tensile strength, thermal conductivity and electrical conductivity and have been proposed as a solid phase extractor for inorganic and organic materials at low concentration levels (Iijima, 1991; Liang et al., 2004; Liang et al., 2005; Seki et al., 2006; Zhou et al., 2006). Among the heavy metal ions tested were cobalt(II), lead(II), cadmium(II), manganese(II), chromium(II) and nickel(II) ions. They found in particular, the reusability of the bacteria immobilised multiwalled carbon nanotubes was as high as 50 cycles without any detrimental loss in sorption capacity. The use of immobilised microbial cells and enzymes coupled with nanotechnology have been reported to improve the reduction of Cr(VI). Application of metal-reducing bacteria with nanomaterials such as siderite have also been reported in the enzymatic Cr(VI) reduction and immobilisation (Seo and Roh, 2015; Gutiérrez-Corona et al., 2016; Fernández et al., 2018). Wang and Chen (2009) spotted another upcoming trend which focuses on the improvement of biomaterials immobilisation, optimization of biosorption process parameters and physicochemical conditions, (including reuse and recycling) where mechanisms involved in biosorption and microorganism-metal interactions can be further studied and elaborated.

Researches in genetic engineering have also been conducted using recombinant microorganisms. A fusion of *E. coli nemA* gene and polyhydroxylalkanoate synthase gene *phaC* from *Ralstonia eutropha* was performed by Robins et al. (2013) in the aims of developing a stable and active chromate reductase. The work produced a microorganism that overexpresses a mutant enzyme (ChrR6) with a chromate reducing efficiency 200-fold greater than the ones found in the wild. The technique appears promising for application in ex situ Cr(VI) remediation. Recent works have also suggested the potential to improve metal uptake by targeting engineered metal binding proteins to the cell surface. Tang et al. (2018) employed CadR, a cadmium specific binding protein, displaying on the surface of *P. aeruginosa* with chromosomal expression and found that the genetically engineered (GE) bacteria showed an adsorption capacity of up to 131.9 μmol/g of Cd(II), where co-existing valence ions in low concentrations had no significant effect on the adsorption capacity. The study points to a direction of using genetically engineered *P. aeruginosa* in heavy metals remediation. Future work may also point in the direction of bacteria cell wall modification to alter the binding of metal ions for biosorption. Several methods have been employed for cell wall modification of microbial and plant-derived biomass in order to enhance the binding capacity of chromium (Thatheyus and Ramya, 2016).

Considerable attention has also been focused on the use of microbial consortia for heavy metals removal (Joutey et al., 2015; Qian et al., 2016; Fernández et al., 2018). Samuel et al. (2013) used a consortium of

Acinetobacter junii, *Escherichia coli*, and *Bacillus subtilis* incubated in a continuous packed bed reactor to improve Cr(VI) removal. Also known as bioaugmentation or the process of adding well-adapted microorganism or microbial consortium to targeted site for remediation, the process is suited for applications where diverse metabolic pathways or robustness is required, typically for higher-scale application.

There has also been much talk about the use of hybrid technologies for pollutants removal (Tsezos, 1999; Wang and Chen, 2009). In the context of bioremediation, hybrid technology involves using a combination of biotechnology-based processes such as biosorption, bioreduction and bioprecipitation and non-biotechnology-based processes such as chemical precipitation, electrochemical processes and membrane technology to treat effluents. The application is analogous to the use of hybrid cars in the automotive industry where a conventional but less sustainable technology (internal combustion engine) is merged with an alternative but more sustainable technology (electric motors) to yield a hybrid system where the advantages of both systems are more holistically exploited. While several examples have been reported (Brady et al., 1994; Thomas and Macaskie, 1996; Riordan et al., 1997; Wang and Chen, 2009), full scale implementation has yet to be seen. What appears to work more efficiently, as advocated as early as 2001, seem to be the design of the hybrid technology encased in one single reactor. Tsezos (1999) suggested the use a hybrid technology (intra-biotechnological or inter-technological), employing a combination of selected processes, most particularly biosorption due to its high efficiency and fast kinetics suited for the treatment of high volume, low concentration effluents. Biosorption may also be used as secondary or polishing applications for heavy metals removal from effluents with low metals concentration employed in conjunction with ion-exchange resin and the also in modular systems for smaller enterprises, such as for specific treatments in targeted applications (Edyvean et al., 1997).

Another area that demands attention lies in the study of biofilms and its relation to heavy metal removal. Biofilm is formed by microorganisms attaching to a surface and then replicating by producing extracellular polymeric substances (EPS). In *P. aeruginosa*, EPS was found to play a significant role in biosorption of metals in biofilms. Various studies have demonstrated the metal sorption ability of *P. aeruginosa* EPS which occurs primarily through complexation of cations with anionic groups on the EPS molecules and metal reductive precipitation (Karthikeyan and Beveridge, 2002; Teitzel and Parsek, 2003; Dogan et al., 2011). Figure 4.2 summarizes the mechanisms in graphical form. Biofilms boast several advantages in biosorption of metals such as high sorption capacities and low production costs. Kurniawan and Yamamoto (2013) first studied the use of natural biofilm as a pollutant ion adsorbent in water contamination treatment. They characterized the natural biofilms formed on the surface of stones from a lake and studied the adsorption kinetics of lithium ions to the biofilm. What they observed was that the biofilm polymer contained both positive and negative charged sites where the adsorption mechanism was driven mainly by electrostatic interactions between metal ions and charged sites. More recently, Kumari et al. (2017) investigated the contribution of immobilised biofilm associated extracellular polymeric substances (EPS) of a marine strain of *P. aeruginosa* N6P6 in sequestering Pb(II). Ca-alginate beads, EPS-alginate beads and alginate-biomass beads (using *P. aeruginosa* N6P6 cell biomass) were compared in terms of capacity to biosorp Pb(II). Calcium alginate was employed to adsorb and support material for immobilisation of EPS and cell biomass. Alginate is a naturally-occurring biopolymer with the ability to form hydrogels with divalent calcium cations via crosslinking. It has several advantages compared to other supporting materials in terms of biodegradability, hydrophilicity, and the presence of carboxylic groups. Results showed that the EPS alginate biosorbent significantly adsorbed Pb(II) followed by biomass alginate biosorbent. The behavior of Pb(II) ion sorption by EPS alginate biosorbent was found to be pH dependent and optimum biosorption of Pb was attained at pH of 6. The work showed the potential use of EPS-alginate as a biosorbent in the remediation of Pb (where the EPS originates from the marine strain of *P. aeruginosa* N6P6).

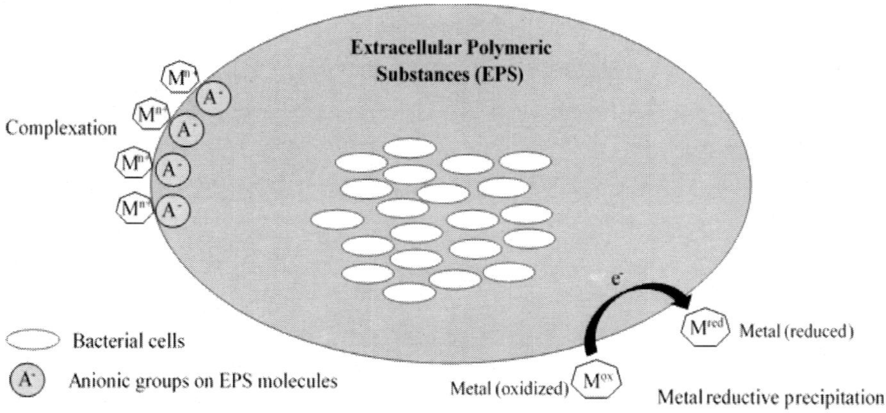

Figure 4.2. Biofilm interaction with metals.

Volesky (2007) pointed out a rather different sort of biosorption, involving the purification and recovery of high-value proteins, steroids and pharmaceuticals drugs such as vinblastine and digoxin, targeted at the product recovery of the high-priced pharmaceuticals. Naik and Dubey (2011) reported the resistance of lead resistant bacteria to multiple drugs. *P. aeruginosa* strain 4EA resistant to lead nitrate, was also shown to be resistant to antibiotics such as ampicillin, erythromycin, amikacin and cephalexin, to mention a few. Similar resistance behavior was also found in another strain *P. aeruginosa* strain WI-1 (Naik et al., 2012). These developments sound alarming as it infers that lead pollution may result in the development of bacterial pathogen resistance to almost all known antibacterial drug or superbugs (Naik and Dubey, 2013).

Recent researches have seen the use of bioremediation to remove specific metals such as selenium. Kora and Rastogi (2016) developed a new method to reduce selenite using *P. aeruginosa* under aerobic conditions. They found that during the process of bacterial conversion, elemental selenium nanoparticles were produced. If properly scaled-up, the bioprocess may be employed as a viable alternative for bioremediation of ground waters and agriculture drainage waters containing selenium. In addition, the green method developed may also be scaled up for industrial scale production of selenium nanoparticles. Kang et al. (2017) examined the toxicity and accumulation behavior of cesium towards *P. aeruginosa*

PAO1 and found that the toxicity of cesium to the bacteria increases as the metal concentration was increased in concentration-dependent manner. The strain was also reported to show high cesium removal efficiency of 76.1% from contaminated water.

CONCLUSION

P. aeruginosa has been shown to be a potential biosorbent in the heavy metal biosorption of waste water systems. The bacteria is versatile and can be applied to a wide range of metals although biosorption mechanisms of the bacteria vary depending on the type of metal encountered. Despite progress made over decades of research work, most biosorption processes involving the bacteria are still conducted at the laboratory scale. Difficulties in commercialisation and complexities in real wastewater matrix are among some of the key barriers to industrial scale applications. In efforts to scale up the process, laboratory waste effluents may serve as a workable test bed, to assess the feasibility of applying the process to larger volumes of wastewaters where the intricacies resembling that of actual wastewater matrix may be encountered. Future work points in the direction of improvement of biomaterials immobilisation, optimization of biosorption process parameters and physicochemical conditions as well as the use of recombinant microorganisms and hybrid technologies, among others.

REFERENCES

Abdel-Aty, A. M., Ammar, N. S., Abdel Ghafar, H. H. and Ali, R. K. (2013). Biosorption of cadmium and lead from aqueous solution by fresh water alga Anabaena sphaerica biomass. *Journal of Advanced Research*, 4(4), pp. 367–374.

Abdi, O. and Kazemi, M. (2015). A review study of biosorption of heavy metals and comparison between different biosorbents. *Journal of Materials and Environmental Science*, 6(5), pp. 1386–1399.

Addour, L., Belhocine, D., Boudries, N., Comeau, Y., Pauss, A., and Mameri, N. (1999). Zinc uptake by *Streptomyces rimosus* biomass using a packed-bed column. *Journal of Chemical Technology & Biotechnology*, 74(11), pp. 1089–1095.

Aguilera, S., Aguilar, M., Chávez, M. P., López-Meza, J. E., Pedraza-Reyes, M., Campos-García, J. and Cervantes, C. (2004). Essential residues in the chromate transporter ChrA of *Pseudomonas aeruginosa*. *FEMS Microbiology Letters*, 232(1), pp. 107–112.

Ahalya, N., Ramachandra, T. V. and Kanamadi, R. D. (2003). Biosorption of heavy metals. *Research Journal Of Chemistry And Environment*, 7(4), pp. 71-79.

Ahemad, M. and Malik, A. (2012). Bioaccumulation of heavy metals by zinc resistant bacteria isolated from agricultural soils irrigated with wastewater. *Bacteriology Journal*, 2(1), pp. 12–21.

Ajao, A., Adebayo, G. B. and Yakubu, S. E. (2011). Bioremediation of textile industrial effluent using mixed culture of *Pseudomonas aeruginosa* and *Bacillus subtilis* immobilized on agar-agar in a bioreactor. *Journal of Microbiology and Biotechnology Research,* 1(3), pp. 50-56.

Aksu, Z. and Kutsal, T. (1991). A bioseparation process for removing lead(II) ions from waste water by using *C. vulgaris*. *Journal of Chemical Technology & Biotechnology*, 52(1), pp. 109–118.

Aksu, Z., Sag, Y. and Kutsal, T. (1992). The biosorpnon of copperod by *C. vulgaris* and *Z. ramigera*. *Environmental Technology*, 13(6), pp. 579–586.

Antunes, W. M., Luna, A. S., Henriques, C. A. and da Costa, A. C. A. (2003). An evaluation of copper biosorption by a brown seaweed under optimized conditions. *Electronic Journal of Biotechnology*, 6(3).

Awasthi, G., Chester, A., Chaturvedi, R. and Prakash, J. (2015). Study on role of *Pseudomonas aeruginosa* on heavy metal bioremediation. *International Journal of Pure & Applied Bioscience,* 3(4), pp. 92–100.

Bakiyaraj, R., Baskaran, L., Chidambaram, A. L., Mahakavi, T. and Santhoshkumar, M. (2014). Bioremediation of chromium by *Bacillus subtilis* and *Pseudomonas aeruginosa. International Journal of Current Microbiology and Applied Sciences*, 3(9), pp. 715–719.

Banjerdkij, P., Vattanaviboon, P. and Mongkolsuk, S. (2005). Exposure to cadmium elevates expression of genes in the OxyR and OhrR regulons and induces cross-resistance to peroxide killing treatment in *Xanthomonas campestris. Applied and Environmental Microbiology*, 71(4), pp. 1843–1849.

Basu, S., Dasgupta, M. and Chakraborty, B. (2014). Removal of chromium (VI) by *Bacillus subtilis* isolated from East Calcutta Wetlands, West Bengal, India. *International Journal of Bioscience, Biochemistry and Bioinformatics*, 4(1), pp. 7–10.

Blindauer, C. A., Harrison, M. D., Robinson, A. K., Parkinson, J. A., Bowness, P. W., Sadler, P. J. and Robinson, N. J. (2002). Multiple bacteria encode metallothioneins and SmtA-like zinc fingers. *Molecular Microbiology*, 45(5), pp. 1421–1432.

Bojórquez, C., Frías Espericueta, M. G. and Voltolina, D. (2016). Removal of cadmium and lead by adapted strains of *Pseudomonas aeruginosa* and *Enterobacter cloacae. Revista Internacional de Contaminación Ambiental*, 32(4), pp. 407–412.

Bouby, M., Billard, I. and MacCordick, J. (1998a). Complexation of Th(VI) with the siderophore pyoverdine A. *Journal of Alloys and Compounds,* 271-273, pp. 206-210.

Bouby, M. Billard, I., MacCordick, J. and Rossini, I.. (1998b). Complexation of U(VI) with the siderophore pyoverdine. *RadiochimicA Acta*, 80(2), p. 95.

Brady, D., Rose, P. D. and Duncan, J. R. (1994). The use of hollow fiber cross-flow microfiltration in bioaccumulation and continuous removal of heavy metals from solution by *Saccharomyces cerevisiae. Biotechnology and Bioengineering*, 44(11), pp. 1362–1366.

Chang, J. S. and Hong, J. (1994). Biosorption of mercury by the inactivated cells of *Pseudomonas aeruginosa* PU21 (Rip64). *Biotechnology and Bioengineering*, 44(8), pp. 999–1006.

Chang, J. S., Law, R. and Chang, C. C. (1997). Biosorption of lead, copper and cadmium by biomass of *Pseudomonas aeruginosa* PU21. *Water Research*, 31(7), pp. 1651–1658.

Chen, X. C., Wang, Y. P., Lin, Q., Shi, J. Y., Wu, W. X. and Chen, Y. X. (2005). Biosorption of copper(II) and zinc(II) from aqueous solution by *Pseudomonas putida* CZ1. *Colloids and Surfaces B: Biointerfaces*, 46(2), pp. 101–107.

Cheung, K. H., Lai, H. Y. and Gu, J. D. (2006). Membrane-associated hexavalent chromium reductase of *Bacillus megaterium* TKW3 with induced expression. *Journal of Microbiology and Biotechnology*, 16(6), pp. 855–862.

Choudhary, S. and Sar, P. (2011). Uranium biomineralization by a metal resistant *Pseudomonas aeruginosa* strain isolated from contaminated mine waste. *Journal of Hazardous Materials*, 186(1), pp. 336–343.

Chovanová, K., Sládeková, D., Kmeť, V., Prokšová, M., Harichová, J., Puškárová, A., Polek, B. and Ferianc, P. (2004). Identification and characterization of eight cadmium resistant bacterial isolates from a cadmium-contaminated sewage sludge. *Biologia Bratislava*, 59(6), pp. 817–827.

Cui, J., Liu, T., Li, F., Yi, J., Liu, C. and Yu, H. (2017). Silica nanoparticles alleviate cadmium toxicity in rice cells: mechanisms and size effects. *Environmental Pollution*, 228, pp. 363–369.

Das, N. (2012). Remediation of radionuclide pollutants through biosorption - an overview. *Clean - Soil, Air, Water*, 40(1), pp. 16–23.

Das, N. and Das, D. (2013). Recovery of rare earth metals through biosorption: An overview. *Journal of Rare Earths*, 31(10), pp. 933–943.

Das, N., Vimala, R. and Karthika, P. (2008). Biosorption of heavy metals - An overview. *Indian Journal of Biotechnology*, 7(2), pp. 159–169.

Desouky, O. A., El-Mougith, A. A., Hassanien, W. A., Awadalla, G. S. and Hussien, S. S. (2016). Extraction of some strategic elements from thorium–uranium concentrate using bioproducts of *Aspergillus ficuum* and *Pseudomonas aeruginosa*. *Arabian Journal of Chemistry*, 9, pp. S795–S805.

Dhankhar, R. and Hooda, A. (2011). Fungal biosorption – an alternative to meet the challenges of heavy metal pollution in aqueous solutions. *Environmental Technology*, 32(5), pp. 467–491.

Dogan, N. M., Kantar, C., Gulcan, S., Dodge, C. J., Yilmaz, B. C. and Mazmanci, M. A. (2011). Chromium (VI) bioremoval by *Pseudomonas* bacteria: role of microbial exudates for natural attenuation and biotreatment of Cr(VI) contamination. *Environmental Science & Technology*, 45, pp. 2278-2285.

Eccles, H. (1999). Treatment of metal-contaminated wastes: why select a biological process? *Trends in Biotechnology*, 17(12), pp. 462–465.

Edyvean, R. G. J., Williams, C. J., Wilson, M. W. and Aderhold., D. (1997). Biosorption using unusual biomasses, in Wase, J. and Forster, C. (eds) *Biosorbents for metal ions*. London, United Kingdom: CRC Press, pp. 165–182.

Eisenbud, M. and Gesell, T. (1997). Natural radioactivity. *Environmental Radioactivity*, pp. 134–200.

Fein, J. B., Fowle, D. A., Cahill, J., Kemner, K., Boyanov, M. and Bunker, B. (2002). Nonmetabolic reduction of Cr(VI) by bacterial surfaces under nutrient-absent conditions. *Geomicrobiology Journal*, 19(3), pp. 369–382.

Fernández, P. M., Viñarta, S. C., Bernal, A. R., Cruz, E. L. and Figueroa, L. I. C. (2018). Bioremediation strategies for chromium removal: Current research, scale-up approach and future perspectives. *Chemosphere*, 208, pp. 139–148.

Fomina, M. and Gadd, G. M. (2014). Biosorption: Current perspectives on concept, definition and application. *Bioresource Technolog,* 160, pp. 3–14.

Fowler, B. A. (1998). Roles of lead-binding proteins in mediating lead bioavailability. *Environmental health perspectives*, 106(Suppl 6), pp. 1585–1587.

Fu, F. and Wang, Q. (2011). Removal of heavy metal ions from wastewaters: A review. *Journal of Environmental Management*, 92(3), pp. 407–418.

Gabr, R. M., Hassan, S. H. A. and Shoreit, A. A. M. (2008). Biosorption of lead and nickel by living and non-living cells of *Pseudomonas aeruginosa* ASU 6a. *International Biodeterioration and Biodegradation*, 62(2), pp. 195–203.

Gadd, G. M. (2009). Biosorption: Critical review of scientific rationale, environmental importance and significance for pollution treatment. *Journal of Chemical Technology and Biotechnology*, 84(1), pp. 13–28.

Gadd, G. M. and White, C. (1985). Copper uptake by *Penicillium ochro-chloron*: influence of pH on toxicity and demonstration of energy-dependent copper influx using protoplasts. *Microbiology*, 131(8), pp. 1875–1879.

Gadd, G. M. and White, C. (1993). Microbial treatment of metal pollution-a working biotechnology? *Trends in Biotechnology*, 11(8), pp. 353–359.

Gadd, G. M., White, C. and de Rome, L. (1988). Heavy metal and radionuclide uptake by fungi and yeasts, in Norris, P. R. and Kelly, D. P. (eds) *Biohydrometallurgy*. Kew, Surrey, UK: Science and Technology Letters, pp. 421-435.

Galun, M., Galun, E., Siegel, B. Z., Keller, P., Lehr, H. and Siegel S. M. (1987). Removal of metal ions from aqueous solutions by Penicillium biomass: kinetic and uptake parameters. *Water, Air, and Soil Pollution*, 33(3–4), pp. 359–371.

Ganguli, A. and Tripathi, A. K. (2002). Bioremediation of toxic chromium from electroplating effluent by chromate-reducing *Pseudomonas aeruginosa* A2Chr in two bioreactors. *Applied Microbiology and Biotechnology*, 58(3), pp. 416–420.

Garbisu, C., Alkorta, I., Llama, M. J. and Serra, J. L. (1998). Aerobic chromate reduction by *Bacillus subtilis*. *Biodegradation*, 9(2), pp. 133–141.

Garnham, G. W. (2003). The use of algae as metal biosorbents', in Wase, J. and Forster, C. (eds.) *Biosorbents for metal ions*. London: Taylor & Francis, pp. 11–33.

Godlewska-Żyłkiewicz, B. and Kozłowska, M. (2005). Solid phase extraction using immobilized yeast *Saccharomyces cerevisiae* for determination of palladium in road dust. *Analytica Chimica Acta*, 539(1-2), pp. 61-67.

Gopalan, R. and Veeramani, H. (1994). Development of a *Pseudomonas sp.* for aerobic chromate reduction. *Biotechnology Techniques*, 8(7), pp. 521–524.

Gupta, V. K., Shrivastava, A. K. and Jain, N. (2001). Biosorption of chromium(VI) from aqueous solutions by green algae *Spirogyra* species. *Water Research*, 35(17), pp. 4079–4085.

Gutiérrez-Corona, J. F., Romo-Rodríguez, P., Santos-Escobar, F., Espino-Saldaña, A. E. and Hernández-Escoto, H. (2016). Microbial interactions with chromium: basic biological processes and applications in environmental biotechnology. *World Journal of Microbiology and Biotechnology*, 32(12), p. 191.

Gvozdyak, P. I., Mogilevich, N. F., Ryl'skii, A. F. and Grishchenko, N. I. (1986). Reduction of hexavalent chromium by collection strains of bacteria. *Mikrobiologiya*, 55, pp. 962–965.

Hansda, A., Kumar, V. and Anshumali (2015). Biosorption of copper by bacterial adsorbents: A Review. *Research Journal of Environmental Toxicology*, 9, pp. 45-58.

Haque, N., Hughes, A., Lim, S. and Vernon, C. (2014). Rare earth elements: overview of mining, mineralogy, uses, sustainability and environmental impact. *Resources*, 3(4), pp. 614–635.

Hassen, A., Saidi, N., Cherif, M. and Boudabous, A. (1998). Effects of heavy metals on *Pseudomonas aeruginosa* and *Bacillus thuringiensis*. *Bioresource Technology*, 65(1–2), pp. 73–82.

Hayes, K. and Leckie, J. O. (1987). Modeling ionic strength effects on cation adsorption at hydrous oxide/solution interfaces. *Journal of Colloid and Interface Science*, 115(2), pp. 564–572.

Herrero, R., Lodeiro, P., Rey-Castro, C., Vilariño, T. and de Vicente, M. E. S. (2005). Removal of inorganic mercury from aqueous solutions by biomass of the marine macroalga *Cystoseira baccata*. *Water Research*, 39(14), pp. 3199–3210.

Horikoshi, T., Nakajima, A. and Sakaguchi, T. (1981). Studies on the accumulation of heavy metal elements in biological systems. *European Journal of Applied Microbiology and Biotechnology*, 12(2), pp. 90–96.

Hu, M. Z. C., Norman, J. M., Faison, B. D. and Reeves, M. E. (1996). Biosorption of uranium by *Pseudomonas aeruginosa* strain CSU : characterization and comparison studies. *Biotechnology and Bioengineering,* 51(2), pp. 237-247.

Huang, G., Zhang, H., Shi, J. X. and Langrish, T. A. G. (2009). Adsorption of chromium(VI) from aqueous solutions using cross-linked magnetic chitosan beads. *Industrial & Engineering Chemistry Research*, 48(5), pp. 2646–2651.

Huang, X., Zhang, G., Pan, A., Chen, F. and Zheng, C. (2016). Protecting the environment and public health from rare earths mining. *Earth's Future*, 4, pp. 532–535.

Husseiny, M. I., El-Aziz, M. A., Badr, Y. and Mahmoud, M. A. (2007). Biosynthesis of gold nanoparticles using *Pseudomonas aeruginosa. Spectrochimica Acta - Part A: Molecular and Biomolecular Spectroscopy*, 67(3–4), pp. 1003–1006.

Iijima, S. (1991). Helical microtubules of graphitic carbon. *Nature*, 354(6348), pp. 56–58.

Ilhan, S., Cabuk, A., Filik, C. and Caliskan, F. (2004). Effect of pretreatment on biosorption of heavy metals by fungal biomass. *Trakya University Journal of Science*, 5(1), pp. 11-17.

Izrael-Živković, L., Rikalović, M., Gojgić-Cvijović, G., Kazazić, S., Vrvić, M., Brčeski, I., r Beškoski, V., Lončarević, B., Gopčević, K. and Karadžić, I. (2018). Cadmium specific proteomic responses of a highly resistant: *Pseudomonas aeruginosa* san ai. *RSC Advances*, 8(19), pp. 10549–10560.

Jobby, R., Jha, P., Yadav, A. K. and Desai, N. (2018). Biosorption and biotransformation of hexavalent chromium [Cr(VI)]: A comprehensive review. *Chemosphere*, 207, pp. 255–266.

Joo, J. H., Hassan, S. H. A. and Oh, S. E. (2010). Comparative study of biosorption of Zn^{2+} by *Pseudomonas aeruginosa* and *Bacillus cereus.*

International Biodeterioration and Biodegradation, 64(8), pp. 734–741.

Joutey, N. T., Sayel, H., Bahafid, W. and El Ghachtoul, N. (2015). Mechanisms of hexavalent chromium resistance and removal by microorganisms. *Reviews of Environmental Contamination and Toxicology*, 233, pp. 45–69.

Kaduková, J. and Virčíková, E. (2005). Comparison of differences between copper bioaccumulation and biosorption. *Environment International*, 31(2), pp. 227–232.

Kamaludeen, S. P. B., Arunkumar, K. R., Avudainayagam, S. and Ramasamy, K. (2003). Bioremediation of chromium contaminated environments. *Indian Journal of Experimental Biology,* 41, pp. 972–985.

Kang, S. Y., Lee, J. U. and Kim, K. W. (2004). Competitive adsorption characteristics of Co^{2+}, Ni^{2+}, and Cr^{3+} by IRN-77 cation exchange resin in synthesized wastewater. *Chemosphere*, 56(2), pp. 141-147.

Kang, S. Y., Lee, J. U. and Kim, K. W. (2007). Biosorption of Cr(III) and Cr(VI) onto the cell surface of *Pseudomonas aeruginosa*. *Biochemical Engineering Journal*, 36(1), pp. 54–58.

Karimpour, M., Ashrafi, S. D., Taghavi, K., Mojtahedi, A., Rookbakhsh, E. and Naghipour, D. (2018). Adsorption of cadmium and lead onto live and dead cell mass of *Pseudomonas aeruginosa*: A dataset. *Data in Brief*, 18, pp. 1185–1192.

Karthikeyan, S. and Beveridge, T. J. (2002). *Pseudomonas aeruginosa* biofilms react with and precipitate toxic soluble gold. *Environmental Microbiology*, 4(11), pp. 667-675.

Kazy, S. K., Sar, P. and D'Souza, S. F. (2008). Studies on uranium removal by the extracellular polysaccharide of a *Pseudomonas aeruginosa* strain. *Bioremediation Journal*, 12(2), pp. 47–57.

Kelly, D. P. (1988). Evolution of the understanding of the microbiology and biochemistry of the mineral leaching habitat, in Norris, P. R. and Kelly, D. P. (eds) *Biohydrometallurgy*. Kew, Surrey, UK: Science and Technology Letters, pp. 3–14.

Kermani, A. J. N., Ghasemi, M. F., Khosravan, A., Farahmand, A. and Shakibaie, M. R. (2010). Cadmium bioremediation by metal-resistant mutated bacteria isolated from active sludge of industrial effluent. *Iranian Journal of Environmental Health Science & Engineering*, 7(4), pp. 279–286.

King, P., Rakesh, N., Beenalahari, S., Prasanna, K. Y. and Prasad, V. S. (2007). Removal of lead from aqueous solution using *Syzygium cumini* L.: Equilibrium and kinetic studies. *Journal of Hazardous Materials*, 142(1–2), pp. 340–347.

Kiran, B., Kaushik, A. and Kaushik, C. P. (2007). Biosorption of Cr(VI) by native isolate of *Lyngbya putealis* (HH-15) in the presence of salts. *Journal of Hazardous Materials*, 141(3), pp. 662–667.

Koçberber, N. and Dönmez, G. (2007). Chromium(VI) bioaccumulation capacities of adapted mixed cultures isolated from industrial saline wastewaters. *Bioresource Technology*, 98(11), pp. 2178–2183.

Konig-Péter, A., Kocsis, B., Kilár, F. and Pernyeszi, T. (2014). Bioadsorption characteristics of *Pseudomonas aeruginosa* PAO1. *Journal of the Serbian Chemical Society*, 79(4), pp. 495–508.

Kora, A. J. and Rastogi, L. (2016). Biomimetic synthesis of selenium nanoparticles by *Pseudomonas aeruginosa* ATCC 27853: An approach for conversion of selenite. *Journal of Environmental Management*, 181, pp. 231–236.

Kotaś, J. and Stasicka, Z. (2000). Chromium occurrence in the environment and methods of its speciation. *Environmental Pollution*, 107(3), pp. 263–283.

Kotrba, P. (2011). Microbial biosorption of metals - general introduction, in Kotrba, P., Mackova, M., and Macek, T. (eds) *Microbial Biosorption of Metals*. Dordrecht: Springer Netherlands, pp. 1-6.

Ku, Y. and Jung, I. L. (2001). Photocatalytic reduction of Cr(VI) in aqueous solutions by UV irradiation with the presence of titanium dioxide. *Water Research*, 35(1), pp. 135-142.

Kumari, S., Mahapatra, S. and Das, S. (2017). Ca-alginate as a support matrix for Pb(II) biosorption with immobilized biofilm associated extracellular polymeric substances of *Pseudomonas aeruginosa* N6P6. *Chemical Engineering Journal*, 328, pp. 556–566.

Kurniawan, A. and Yamamoto, T. (2013). Biofilm polymer for biosorption of pollutant ions. *Procedia Environmental Sciences*, 17, pp. 179–187.

Lam, T. V. Agovina, P., Niu, X. and Roché, L. (2007). Linkage study of cancer risk among lead-exposed workers in New Jersey. *Science of the Total Environment*, 372(2–3), pp. 455–62.

Latha, A. and Reddy, S. (2013). Review on bioremediation – potential tool for removing environmental pollution. *International Journal of Basic and Applied Chemical Sciences*, 3(3), pp. 21-33.

Leiterer, A., Berard, P. and Menetrier, F. (2010). Thorium and health: state of the art. *International Nuclear Information System Report,* 42(39).

Levinaa, A., Mulyania, I. and Lay, P. A. (2007). Redox chemistry and biological activities of chromium(III) complexes, in Vincent, J. B. (ed.) *The Nutritional Biochemistry of Chromium (III).* Amsterdam, Netherlands: Elsevier, pp. 225–256.

Liang, P., Liu, Y., Guo, L., Zeng, J. and Lu, H. (2004). Multiwalled carbon nanotubes as solid-phase extraction adsorbent for the preconcentration of trace metal ions and their determination by inductively coupled plasma atomic emission spectrometry. *Journal of Analytical Atomic Spectrometry*, 19(11), pp. 1489-1492.

Liang, P., Ding, Q. and Song, F. (2005). Application of multiwalled carbon nanotubes as solid phase extraction sorbent for preconcentration of trace copper in water samples. *Journal of Separation Science*, 28(17), pp. 2339–2343.

Liu, T., Nakashima, S., Hirose, K., Uemura, Y., Shibasaka, M., Katsuhara, M. and Kasamo, K. (2003). A metallothionein and CPx-ATPase handle heavy-metal tolerance in the filamentous cyanobacterium *Oscillatoria brevis*. *FEBS Letters*, 542(1–3), pp. 159–163.

Liu, Y. G., Xu, W. H., Zeng, G. M., Li, X. and Gao, H. (2006). Cr(VI) reduction by *Bacillus sp*. isolated from chromium landfill. *Process Biochemistry*, 41(9), pp. 1981–1986.

Llovera, S., Bonet, R., Simon-Pujol, M. D. and Congregado, F. (1993). Chromate reduction by resting cells of *Agrobacterium radiobacter*

EPS-916. *Applied and Environmental Microbiology*, 59(10), pp. 3516–3518.

Lovley, D. R. (1993). Dissimilatory metal reduction. *Annual Review of Microbiology*, 47(1), pp. 263–290.

MacHalová, L., Pipíška, M., Trajteaová, Z. and Horník, M. (2015). Comparison of Cd^{2+} biosorption and bioaccumulation by bacteria-a radiometric study. *Nova Biotechnologica et Chimica*, 14(2), pp. 158–175.

Mack, C., Wilhelmi, B., Duncan, J. R. and Burgess, J. E. (2007). Biosorption of precious metals. *Biotechnology Advances*, 25(3), pp. 264–271.

Malik, A. (2004). Metal bioremediation through growing cells. *Environment International*, 30(2), pp. 261–278.

Martinez, R. J., Beazley, M. J., Taillefert, M., Arakaki, A. K., Skolnick, J. and Sobecky, P. A. (2007). Aerobic uranium (VI) bioprecipitation by metal-resistant bacteria isolated from radionuclide- and metal-contaminated subsurface soils. *Environmental Microbiology*, 9(12), pp. 3122–3133.

Mathiyazhagan, N. and Natarajan, D. (2011). Bioremediation on effluents from magnesite and bauxite mines using *Thiobacillus Spp* and *Pseudomonas Spp*. *Journal of Bioremediation & Biodegradation*, 2(1), pp. 2–7.

McLean, J. S., Beveridge, T. J. and Phipps, D. (2000). Isolation and characterization of a chromium-reducing bacterium from a chromated copper arsenate-contaminated site. *Environmental Microbiology*, 2(6), pp. 611–619.

Mitra, R. S., Gray, R. H., Chin, B. and Bernstein, I. A. (1975). Molecular mechanisms of accommodation in *Escherichia coli* to toxic levels of Cd^{2+}. *Journal of Bacteriology*, 121(3), pp. 1180–1188.

Mullen, M. D., Wolf, D. C., Ferris, F. G., Beveridge, T. J., Flemming, C. A. and Bailey, G. W. (1989). Bacterial sorption of heavy-metals. *Applied and Environmental Microbiology*, 55(12), pp. 3143–3149.

Muneer, B., Iqbal, M. J., Shakoori, F. R. and Shakoori, A. R. (2016). Isolation, identification and cadmium processing of *Pseudomonas aeruginosa* (EP-Cd1) isolated from soil contaminated with electroplating industrial wastewater. *Pakistan Journal of Zoology*, 48(5), pp. 1495–1501.

Munter, R. (2013). Technology for the removal of radionuclides from natural water and waste management : state of the art. *Proceedings of the Estonian Academy of Sciences*, 62(2), pp. 122–132.

Naik, M. M. and Dubey, S. K. (2011). Lead-enhanced siderophore production and alteration in cell morphology in a Pb-resistant *Pseudomonas aeruginosa* strain 4EA. *Current Microbiology*, 62(2), pp. 409–414.

Naik, M. M. and Dubey, S. K. (2013). Lead resistant bacteria: Lead resistance mechanisms, their applications in lead bioremediation and biomonitoring. *Ecotoxicology and Environmental Safety*. 98, pp. 1–7.

Naik, M. M., Pandey, A. and Dubey, S. K. (2012). *Pseudomonas aeruginosa* strain WI-1 from Mandovi estuary possesses metallothionein to alleviate lead toxicity and promotes plant growth. *Ecotoxicology and Environmental Safety*, 79, pp. 129–133.

Naja, G. and Volesky, B. (2011). The mechanism of metal cation and anion biosorption, in Kotrba, P., Mackova, M., and Macek, T. (eds) *Microbial Biosorption of Metals*. Dordrecht: Springer Netherlands, pp. 19–58.

Nakajima, A. and Sakaguchi, T. (1986). Selective accumulation of heavy metals by microorganisms. *Applied Microbiology and Biotechnology*, 24(1), pp. 59–64.

Naro, F., Tordi, M. G., Giacometti, G. M., Tomei, F., Timperio, A. M. and Zolla, L. (2000). Metal binding to *Pseudomonas aeruginosa* azurin: A kinetic investigation. *Zeitschrift fur Naturforschung - Section C Journal of Biosciences*, 55(5–6), pp. 347–354.

Nies, D. H. (2003). Efflux-mediated heavy metal resistance in prokaryotes. *FEMS Microbiology Reviews*, 27(2–3), pp. 313–339.

Norberg, A. B. and Persson, H. (1984). Accumulation of heavy-metal ions by *Zoogloea ramigera*. *Biotechnology and Bioengineering*, 26(3), pp. 239–246.

Norberg, A. and Rydin, S. (1984). Development of a continuous process for metal accumulation by *Zoogloea ramigera*. *Biotechnology and Bioengineering*, 26(3), pp. 265–268.

Novick, R. P. and Roth, C. (1968). Plasmid-linked resistance to inorganic salts in *Staphylococcus aureus*. *Journal of Bacteriology*, 95(4), pp. 1335–1342.

Oaikhena, E. E., Makaije, D. B., Denwe, S. D., Namadi, M. M. and Haroun, A. A. (2016). Bioremediation potentials of heavy metal tolerant bacteria isolated from petroleum refinery effluent. *American Journal of Environmental Protection*, 5(2), pp. 29–34.

Ojiagu, K. D., Odibo, F. J. C., Ojiagu, N. C., Agu, K. C. and Okafor, A. C. (2018). Biosorption of hexavalent chromium by *Pseudomonas aeruginosa* strain ANSC: equilibria isothermic, kinetic and thermodynamic studies. *Bioengineering and Bioscience*, 6(1), pp. 1–10.

Ojovan, M. I. and Lee, W. E. (2014). An introduction to nuclear waste immobilisation, second edition. Kidlington, Oxford, U.K.: Elsevier.

Pagano, G., Guida, M., Tommasi, F. and Oral, R. (2015). Health effects and toxicity mechanisms of rare earth elements - knowledge gaps and research prospects. *Ecotoxicology and Environmental Safety*, 115, pp. 40–48.

Pardo, R., Herguedas, M., Barrado, E., and Vega, M. (2003). Biosorption of cadmium, copper, lead and zinc by inactive biomass of *Pseudomonas putida*. *Analytical and Bioanalytical Chemistry*, 376, pp. 26-32.

Park, C. H., Gonzales, C., Ackerley, D., Keyhan, M. and Matin, A. (2002). Molecular engineering of soluble bacterial proteins with chromate reductase activity, in Pellei, M., Porta, A., and Hinchee, R. E. (eds) *Remediation and beneficial reuse of contaminated sediments*. Columbus: Batelle Press, pp. 103-111.

Park, D., Yun, Y. S., Cho, H. Y. and Park, J. M. (2004). Chromium biosorption by thermally treated biomass of the brown seaweed, *Ecklonia* sp.. *Industrial & Engineering Chemistry Research*, 43(26), pp. 8226–8232.

Park, D., Yun, Y. S., Yim, K. H. and Park, J. M. (2006). Effect of Ni(II) on the reduction of Cr(VI) by *Ecklonia* biomass. *Bioresource Technology*, 97(14), pp. 1592–1598.

Park, D., Yun, Y. S. and Park, J. M. (2010). The past, present, and future trends of biosorption. *Biotechnology and Bioprocess Engineering*, 15(1), pp. 86–102.

Park, R. M., Bena, J. F., Stayner, L. T., Smith, R. J., Gibb, H. J. and Lees, P. S. J. (2004). Hexavalent chromium and lung cancer in the chromate industry: a quantitative risk assessment. *Risk Analysis*, 24(5), pp. 1099–1108.

Pérez Silva, R. M., Ábalos Rodríguez,, A., Gómez Montes De Oca, J. M. and Cantero Moreno, D. (2009). Biosorption of chromium, copper, manganese and zinc by *Pseudomonas aeruginosa* AT18 isolated from a site contaminated with petroleum. *Bioresource Technology*, 100(4), pp. 1533–1538.

Philip, L., Iyengar, L. and Venkobachar, C. (2000). Biosorption of U, La, Pr, Nd, Eu and Dy by *Pseudomonas aeruginosa*. *Journal of Industrial Microbiology and Biotechnology*, 25(1), pp. 1–7.

Qian, J., Wei, L., Liu, R., Jiang, F., Hao, X. and Chen, G. (2016). An exploratory study on the pathways of Cr (VI) reduction in sulfate-reducing up-flow anaerobic sludge bed (UASB) reactor. *Scientific Reports*, 6(1), p. 23694.

Raabe, O. G. (2010). Concerning the health effects of internally deposited radionuclides. *Health Physics*, 98(3), pp. 515–536.

Raghuraman, T. Geoffrey, C. J., Suriyanarayanan, S. and Thatheyus, A. J. (2013). Chromium removal by using chosen Pseudomonads. *American Journal of Environmental Protection*, 1(1), pp. 14–16.

Rasmey, A. H., Aboseidah, A. A. and Youssef, A. K. (2018). Application of Langmuir and Freundlich isotherm models on biosorption of Pb^{2+} by freez-dried biomass of *Pseudomonas aeruginosa*. *Egypt Journal of Microbiology*, 53, pp. 37-48.

Rim, K. T. (2015). Effects of rare earth elements on the environment and human health: A literature review. *Toxicology and Environmental Health Sciences*, 8(3), pp. 189–200.

Riordan, C. *et al.* (1997) 'Removal of uranium from solution using residual brewery yeast: Combined biosorption and precipitation', *Biotechnology Letters*, 19(4), pp. 385–387.

Robins, K. J., Hooks, D. O., Rehm, B. H. A. and Ackerley, D. F. (2013). *Escherichia coli* NemA is an efficient chromate reductase that can be biologically immobilized to provide a cell free system for remediation of hexavalent chromium. *PLoS ONE*, 8(3), pp. e59200.

Rodríguez, C. E. and Quesada, A. (2006). Nickel biosorption by *Acinetobacter baumannii* and *Pseudomonas aeruginosa* isolated from industrial wastewater. *Brazilian Journal of Microbiology*, 37(4), pp. 465–467.

Romanenko, V. I. and Koren'kov, V. N. (1977). Pure culture of bacteria using chromates and bichromates as hydrogen acceptors during development under anaerobic conditions. *Mikrobiologiia*, 46(3), pp. 414–417.

Salbu, B., Skipperud, L. and Lind, O. C. (2015). Sources contributing to radionuclides in the environment: with focus on radioactive particles', in Walther, C. and Gupta, D. K. (eds) *Radionuclides in the Environment*. Cham: Springer International Publishing, pp. 1–36.

Samuel, J., Pulimi, M., Paul, M. L., Maurya, A., Chandrasekaran, N. and Mukherjee, A. (2013). Batch and continuous flow studies of adsorptive removal of Cr(VI) by adapted bacterial consortia immobilized in alginate beads. *Bioresource Technology*, 128, pp. 423–430.

Sar, P., Kazy, S. K., Asthana, R. K. and Singh, S. P. (1999). Metal adsorption and desorption by lyophilized *Pseudomonas aeruginosa*. *International Biodeterioration & Biodegradation*, 44(2–3), pp. 101–110.

Schiewer, S. and Volesky, B. (1996). Modeling multi-metal ion exchange in biosorption. *Environmental Science & Technology*, 30(10), pp. 2921–2927.

Schiewer, S. and Volesky, B. (1997). Ionic strength and electrostatic effects in biosorption of divalent metal ions and protons. *Environmental Science & Technology*, 31(9), pp. 2478–2485.

Seki, H., Noguchi, A., Suzuki, A. and Inoue, N. (2006). Biosorption of heavy metals onto Gram-positive bacteria, *Lactobacillus plantarum* and *Micrococcus luteus*. *Kagaku Kogaku Ronbunshu*, 32(4), pp. 352–355.

Seki, H., Suzuki, A. and Mitsueda, S. (1998). Biosorption of heavy metal ions on *Rhodobacter sphaeroides* and *Alcaligenes eutrophus* H16. *Journal of Colloid and Interface Science*, 197(2), pp. 185–190.

Seo, H. and Roh, Y. (2015). Biotransformation and its application: biogenic nano-catalyst and metal-reducing-bacteria for remediation of Cr(VI)-contaminated water. *Journal of Nanoscience and Nanotechnology*, 15(8), pp. 5649–5652.

Shahid, M., Pinelli, E. and Dumat, C. (2012). Review of Pb availability and toxicity to plants in relation with metal speciation; role of synthetic and natural organic ligands. *Journal of Hazardous Materials*, 219–220, pp. 1–12.

Sheng, P. X., Ting, Y., Chen, J. P. and Hong, L. (2004). Sorption of lead, copper, cadmium, zinc, and nickel by marine algal biomass: characterization of biosorptive capacity and investigation of mechanisms. *Journal of Colloid and Interface Science*, 275(1), pp. 131–141.

Singh, R., Bishnoi, N. R., Kirrolia, A. and Kumar, R. (2013). Synergism of *Pseudomonas aeruginosa* and Fe^0 for treatment of heavy metal contaminated effluents using small scale laboratory reactor. *Bioresource Technology*, 127, pp. 49–58.

Sinha, S. and Mukherjee, S. K. (2009). *Pseudomonas aeruginosa* KUCd1, a possible candidate for cadmium bioremediation. *Brazilian Journal of Microbiology*, 40, pp. 655–662.

Srinath, T., Verma, T., Ramteke, P. W. and Garg, S. K. (2002). Chromium (VI) biosorption and bioaccumulation by chromate resistant bacteria. *Chemosphere*, 48(4), pp. 427–435.

Strandberg, G. W., Shumate, S. E. and Parrot Jr, J. R. (1981). Microbial cells as biosorbents for heavy metals : Accumulation of uranium by *Saccharomyces cerevisiae* and *Pseudomonas aeruginosa*. *Applied and Environmental Microbiology*, 41(1), pp. 237–245.

Straube, W. L., Nestler, C. C., Hansen, L. D., Ringleberg, D., Pritchard P. H. and Jones-Meehan, J. (2003). Remediation of polyaromatic hydrocarbons (PAHs) through landfarming with biostimulation and bioaugmentation. *Acta Biotechnologica*, 23(2-3), pp. 179–196.

Stumm, W. and Morgan, J. J. (1996). *Aquatic chemistry: chemical equilibria and rates in natural waters*. 3rd ed. New York: Wiley.

Suzuki, T., Miyata, N., Horitsu, H., Kawai, K., Takamizawa, K., Tai, Y. and Okazaki. M. (1992). NAD(P)H-dependent chromium (VI) reductase of *Pseudomonas ambigua* G-1: a Cr(V) intermediate is formed during the reduction of Cr(VI) to Cr(III), *Journal of Bacteriology*, 174(16), pp. 5340–5345.

Suzuki, Y. and Banfield, J. F. (2004). Resistance to, and accumulation of, uranium by bacteria from a uranium-contaminated site. *Geomicrobiology Journal*, 21(2), pp. 113–121.

Tang, X., Tang, X., Zeng, G., Fan, C., Zhou, M., Tang, L., Zhu, J., Wan, J., Huang, D., Chen, M., Xu, P., Zhang, C., Lu, Y. and Xiong, W. (2018). Chromosomal expression of CadR on *Pseudomonas aeruginosa* for the removal of Cd(II) from aqueous solutions. *Science of the Total Environment*, 636, pp. 1355–1361.

Teitzel, G. M. and Parsek, M. R. (2003) Heavy metal resistance of biofilm and planktonic *Pseudomonas aeruginosa*. *Applied and Environmental Microbiology*, 69(4), pp. 2313-2320.

Texier, A. C., Andrès, Y. and Le Cloirec, P. (1999). Selective biosorption of lanthanide (La, Eu, Yb) ions by *Pseudomonas aeruginosa*. *Environmental Science & Technology*, 33(3), pp. 489–495.

Texier, A. C., Andrès, Y., Illemassene, M. and Le Cloirec, P. (2000). Characterization of lanthanide ions binding sites in the cell wall of *Pseudomonas aeruginosa*. *Environmental Science & Technology*, 34(4), pp. 610–615.

Texier, A. C., Andrès, Y., Faur-Brasquet, C. and Le Cloirec, P. (2002). Fixed-bed study for lanthanide (La, Eu, Yb) ions removal from aqueous solutions by immobilized *Pseudomonas aeruginosa*: Experimental data and modelization. *Chemosphere*, 47(3), pp. 333–342.

Thatheyus, J. A. and Ramya, D. (2016). Biosorption of chromium using bacteria: An overview. *Science International*, 4, pp. 74–79.

Thomas, R. A. P. and Macaskie, L. E. (1996). Biodegradation of tributyl phosphate by naturally occurring microbial isolates and coupling to the removal of uranium from aqueous solution. *Environmental Science & Technology*, 30(7), pp. 2371–2375.

Tong, S., Von Schirnding, Y. E. and Prapamontol, T. (2000). Environmental lead exposure: a public health problem of global dimensions. *Bulletin of the World Health Organization*, 78(9), pp. 1068–1077.

Tordi, M. G., Naro, F., Giordano, R. and Silvestrini, M. C. (1990). Silver binding to *Pseudomonas aeruginosa* azurin. *Biology of Metals*, 3, pp. 73–76.

Tsezos, M. (1999). Biosorption of metals. The experience accumulated and the outlook for technology development. *Biohydrometallurgy and the Environment Toward the Mining of the 21st Century - Proceedings of the International Biohydrometallurgy Symposium,* pp. 171-173.

Tsuruta, T. (2004). Biosorption and recycling of gold using various microorganisms. *The Journal of General and Applied Microbiology*, 50(4), pp. 221–228.

Turner, J. S., Glands, P. D., Samson, A. C. and Robinson, N. J. (1996). Zn^{2+}-sensing by the cyanobacterial metallothionein repressor SmtB: different motifs mediate metal-induced protein-DNA dissociation. *Nucleic Acids Research*, 24(19), pp. 3714–3721.

Tuzen, M., Saygi, K. O., Usta, C. and Soylak, M. (2008). *Pseudomonas aeruginosa* immobilized multiwalled carbon nanotubes as biosorbent for heavy metal ions. *Bioresource Technology*, 99(6), pp. 1563–1570.

Tuzen, M. and Soylak, M. (2008). Biosorption of aluminum on *Pseudomonas aeruginosa* loaded on Chromosorb 106 prior to its graphite furnace atomic absorption spectrometric determination. *Journal of Hazardous Materials*, 154(1–3), pp. 519–525.

Uslu, G. and Tanyol, M. (2006). Equilibrium and thermodynamic parameters of single and binary mixture biosorption of lead (II) and copper (II) ions onto *Pseudomonas putida*: effect of temperature. *Journal of Hazardous Materials*, 135, pp. 87-93.

Van Gosen, B. S., Verplanck, P. L., Long, K. R., Gambogi, J. and Seal, R. R. (2014). The rare-earth elements-vital to modern technologies and lifestyles. *U.S. Geological Survey Fact Sheet*.

Veglio, F. and Beolchini, F. (1997). Removal of metals by biosorption: A review. *Hydrometallurgy*, 44(3), pp. 301–316.

Vijayaraghavan, K. and Balasubramanian, R. (2015). Is biosorption suitable for decontamination of metal-bearing wastewaters? A critical review on the state-of-the-art of biosorption processes and future directions. *Journal of Environmental Management*, 160, pp. 283–296.

Vijayaraghavan, K. and Joshi, U. M. (2013). Hybrid *Sargassum*-sand sorbent: A novel adsorbent in packed column to treat metal-bearing wastewaters from inductively coupled plasma-optical emission spectrometry. *Journal of Environmental Science and Health, Part A*, 48(13), pp. 1685–1693.

Vijayaraghavan, K. and Yun, Y. S. (2008). Bacterial biosorbents and biosorption. *Biotechnology Advances*, 26(3), pp. 266–291.

Volesky, B. (1990). Removal and recovery of heavy metals by biosorption', in Volesky, B. (ed.) *Biosorption of Heavy Metals*. Boca Raton: CRC Press, pp. 7–43.

Volesky, B. (2007). Biosorption and me. *Water Research*, 41(18), pp. 4017–4029.

Wang, J. and Chen, C. (2009). Biosorbents for heavy metals removal and their future. *Biotechnology Advances*, 27(2), pp. 195–226.

Wang, Y. T. and Xiao, C. (1995). Factors affecting hexavalent chromium reduction in pure cultures of bacteria. *Water Research*, 29(11), pp. 2467–2474.

Watt, G. C., Britton, A., Gilmour, H. G., Moore, M. R., Murray, G. D. and Robertson, S. J. (2000). Public health implications of new guidelines for lead in drinking water: a case study in an area with historically high water lead levels. *Food and Chemical Toxicology*, 38(1 Suppl), pp. S73-79.

Won, S. W., Kotte, P., Wei, W., Lim, A. and Yun, Y. (2014). Biosorbents for recovery of precious metals. *Bioresource Technology*, 160, pp. 203–212.

Xu, W. H., Liu, Y. G., Zeng, G. M., Li, X., Song, H. X. and Peng, Q. Q. (2009). Characterization of Cr(VI) resistance and reduction by *Pseudomonas aeruginosa*. *Transactions of Nonferrous Metals Society of China (English Edition)*, 19(5), pp. 1336–1341.

Xu, X., Stewart, P. S. and Chen, X. (1996). Transport limitation of chlorine disinfection of *Pseudomonas aeruginosa* entrapped in alginate beads. *Biotechnology and Bioengineering*, 49(1), pp. 93–100.

Yadav, R., Ali, M., Kumar, A. and Pandey, B. N. (2017). Mechanism of carcinogenesis after exposure of actinide radionuclides: Emerging concepts and missing links. *Journal of Radiation and Cancer Research*, 8(1), pp. 20–34.

Yin, K., Lv, M., Wang, Q., Wu, Y., Liao, C., Zhang, W. and Chen, L. (2016). Simultaneous bioremediation and biodetection of mercury ion through surface display of carboxylesterase E2 from *Pseudomonas aeruginosa* PA1. *Water Research*, 103, pp. 383–390.

Zeng, X., Tang, J., Liu, X. and Jiang, P. (2009). Isolation, identification and characterization of cadmium-resistant *Pseudomonas aeruginosa* strain E1. *Journal of Central South University of Technology*, 16(3), pp. 416–421.

Zhou, Q., Wang, W., Xiao, J., Wang, J., Liu, G., Zhi, Q. and Guo, G. (2006). Comparison of the enrichment efficiency of multiwalled carbon nanotubes, C18 silica, and activated carbon as the adsorbents for the solid phase extraction of atrazine and simazine in water samples. *Microchimica Acta*, 152(3–4), pp. 215–224.

Zolgharnein, H., Karami, K., Assadi, M. M. and Sohrab, A. D. (2010). Heavy metal biosorption on *Pseudomonas aerugionsa* strain MCCB 102 isolated from the Persian Gulf. *Asian Journal of Biotechnology*, 2(2), pp. 99–109.

In: *Pseudomonas aeruginosa* ISBN: 978-1-53616-100-7
Editors: C. Fah Wong and H. Idris

Chapter 5

PHOTODYNAMIC THERAPY AS A METHOD OF *Pseudomonas aeruginosa* MANAGEMENT IN INFECTED RADIAL SKIN ULCERS

Nikolay V. Krasnoselskiy[1], Larisa I. Simonova[1], Nadezhda E. Sklyar[2] and Vladislava V. Sarkis-Ivanova[2]

[1]State Institution "Grigoriev Institute for Medical Radiology National Academy of Medical Sciences of Ukraine", Kharkiv, Ukraine

[2]State Institution "I. Mechnikov Institute of Microbiology and Immunology National Academy of Medical Sciences of Ukraine", Kharkiv, Ukraine

ABSTRACT

Pseudomonas aeruginosa attracts attention as a causative agent of nosocomial infections, resistant to a wide range of antimicrobial agents. Photodynamic impact on bacteria is an alternative antibiotic therapy for sanitizing infected wounds. *The purpose of this study* was to experimentally study the effectiveness of photodynamic therapy for the treatment of skin ulcers infected by *Pseudomonas aeruginosa*. A 0.1% aqueous solution of methylene blue was chosen as a photosensitizer, the irradiation was performed with LED red light (λ = 630-650 nm). *In vitro* trials showed 100% death of microorganisms in exposure to PDT for 30 minutes. The study of persistent properties of test strains exposed to sublethal photodynamic effects revealed inhibition of their adhesive properties and biofilm-forming activity. *In vivo* trials on rats with infected radiation lesions of the skin confirmed the effectiveness of the developed PDT scheme. An antibiotic-resistant clinical strain of *Pseudomonas aeruginosa* with a high capacity for biofilm formation was used for infection. It has been determined that in order to achieve complete elimination of the etiological factor, it is necessary to conduct two sessions of PDT in 5 days. PDT is widely used in oncology, surgery and other areas of medicine. The results of the study expand the field of application of PDT as an alternative to antimicrobial therapy in the treatment of infected radiation lesions of the skin, including cases with antibiotic-resistant strains of *Pseudomonas aeruginosa* with high biofilm capacity as the etiological factor.

INTRODUCTION

Recently, *Pseudomonas aeruginosa* has attracted attention as one of the main causative agents of nosocomial infections. A clinically important feature of these bacteria is the high incidence of resistance to various classes of antimicrobial agents. Over the recent decade, the percentage of pan-antibiotic-resistant strains has grown rapidly. Another equally important feature of *P. aeruginosa* strains is the extremely high ability to form a biofilm, which structure and physiological properties provide increased resistance to antimicrobial agents and help to overcome the protective forces of the macroorganism in infection (Hirsch, 2010; Bjarnsholt et al., 2010).

Radiation skin ulcers can occur as a result of radiation accidents, military clashes or terrorist acts, and in clinical situations, as side effects of medical radiation for diagnostic or therapeutic purposes. Frequently, radiation damage in peacetime occurs as a complication of radiotherapy (Pylypenko, 2014). Radiation ulcers of the skin can be complicated by the bacterial infection and the development of local, and in some cases, systemic purulent-inflammatory processes. Our previous studies have shown that local irradiation of the skin of laboratory rats causes a qualitative and quantitative increase in microbial colonization of the irradiated area and the emergence of additional types of microorganisms, including pathogens (Simonova-Pushkar et al., 2015).

In-patient treatment of oncologic patients significantly increases the risk of infection of radiation ulcers with antibiotic-resistant nosocomial strains of bacteria. It is known that antibiotic therapy is a traditional method of treatment of infected radiation ulcers. However, such treatment is significantly complicated by the development of microbial resistance to antibiotics. It is the prevalence of antibiotic resistant microflora that stipulates the search for alternative methods of fighting infections, and in this sense photodynamic therapy (PDT) method occupies an increasingly important place (Jori et al., 2006; Winkler et al., 2016). PDT is a light-dependent method and includes two main components, such as light irradiation of different wavelengths in the optical range and a photosensitizer (PS) that can accumulate in target cells (Chatzopoulos and Doufexi, 2016).

The main purpose of the photosensitizer is the ability to absorb the incident energy of the directed light flux with the development of photochemical reaction and the transfer of the generated excitation energy to molecular oxygen present in the tissues, which leads to the formation of a large number of reactive oxygen species (ROS), primarily singlet oxygen ($^1O_2$) and superoxide anion, the oxygen radical (O_2– •). The latter triggers a whole chain of oxidative processes, which lead to the death of target cells through apoptosis or necrosis (Chatzopoulos and Doufexi, 2016).

The method of PDT is widely used in oncology, surgery and other fields of medicine. One of the new directions in the use of PDT is the treatment of infected skin and soft tissue defects (Jori et al., 2006; Stranadko, 2011). For these purposes, various types of photosensitizers are used. Photosensitizer is a special substance capable of selectively accumulating in target cells and subsequently actively influencing them under the influence of visible light with a wavelength corresponding to the absorption maximum of a specific PS. For PDT, a large amount of PS has been developed with various routes of administration: enteral, parenteral, inhalation, topical (Jori et al., 2006). PS is most often used for the treatment of wounds and superficial injuries of the skin and soft tissues. In this respect, some dyes (methylene blue, brilliant green, gentian violet) and phyto-agents (chlorophyllite, novoyimanin) possess photosensitizing properties (Plavskiy et al., 2010; Sadykov et al., 2011).

THE EFFECT OF PHOTODYNAMIC THERAPY ON GROWTH AND BIOFILM FORMING PROPERTIES OF *Pseudomonas aeruginosa* STRAINS (*IN VITRO*)

Antibacterial photodynamic therapy is based on oxidative degradation of microorganisms with a combined effect of photosensitizer and light with a certain wavelength and currently considered as an alternative to antibiotic therapy. Cytotoxic products produced by PDT have a short half-life and a limited radius of effect. Their antimicrobial activity is non-selective, that is, this method can be used to destroy any microbial cells, in particular, antibiotic-resistant ones (Jori et al., 2006; Kashef et al., 2011). This gives PDT an advantage over antibiotic therapy in the possibility of application without first determining the sensitivity of bacteria and absence of resistance development in target microorganisms. The results of our study have shown that PDT can effectively eliminate the effect of the antibiotic-specific etiological factor of purulent-inflammatory complications of radiation skin ulcer.

The first phase of the study was conducted *in vitro* using reference *P. aeruginosa* strain and 9 clinical isolates to select the optimal concentration, exposure time of the selected photosensitizer (methylene blue) and the exposure parameters of irradiation with LED red light (λ = 630-650 nm).

Daily test cultures were washed off from the surface of agar medium with a sterile physiological solution and adjusted to 1.0 units of McFarland's optical density standard. Cultures were synchronized before preparation of samples by a one-time effect of low temperature (4°C) for 30 minutes (Basnakyan, 1992). The produced bacterial suspensions were sequentially ten times diluted to 10^{-3} with a physiological solution. The number of living microorganisms (colony-forming units) was determined by serial dilutions with subsequent inoculation on the appropriate nutrient medium.

Test samples were prepared by putting and evenly distributing 0.1 ml of the prepared suspension of bacterial cells from various dilutions in the prepared Petri dishes with nutrient agar. The dishes inoculated with microorganisms (each concentration separately), were further added 0.2 ml of 0.05% or 0.1% solution of methylene blue (Methylenum coeruleum – $C_{16}H_{18}ClN_3S{\cdot}3H_2O$). Photosensitizer exposure was 30 minutes at room temperature in the dark. The dishes inoculated in a similar manner, but without applying photosensitizer served as controls. Photon unit "Barva-LED/630" was used as a light source in the trials. LED irradiation of the experimental inoculated dishes was performed for 20, 40 or 60 minutes. Experimental and control dishes were placed in a thermostat at a temperature of 37°C for 18-20 hours. After incubation, the number of the grown colonies in the experimental and control groups was counted. The number of microorganisms in the nutrient medium on Petri dishes was expressed in decimal logarithms of the number of microorganisms per milliliter (lg CFU/ml). This is the indicator of the volumetric concentration of colony-forming units (CFUs) per volume unit, expressed in decimal logarithms. In determining the antimicrobial effect of PDT, the study was conducted in triple repetitions.

Adhesive properties of the test strains were identified by studying adhesive potency of microorganisms (Brilis et al., 1986). The study implied commercial formalized erythrocytes from lambs with a concentration of 10^8 cells/ml. For the experiment, 0.5 ml of a suspension of formalized erythrocytes and a suspension of microorganisms with a concentration of 10^9 CFU/ml were introduced into the test tubes. The mixture was incubated at 37°C for 30 minutes, being shaken from time to time. Then, a smear was prepared on a carefully de-oiled slide was prepared, which was dried at room temperature, fixed and stained by

Romanovsky-Gimza method (1904). The following criteria were used to evaluate the adhesive properties of the strain: the average adhesion rate (AAR), adhesion coefficient (AC) and the index of adhesion of microorganisms (IAM). Determination of these indices was carried out on fifty erythrocytes, studying each of them on the object glass. IAM (index of adhesion of microorganisms) is the average number of microbial cells attached to one erythrocyte, which participates in the adhesive process. This index was calculated according to the formula (Figure 5.1).

$$IAM = \frac{AAR \times 100}{AC}$$

Figure 5.1. The formula for calculating the index of adhesion of microorganisms where **IAM** is the index of adhesion of microorganisms; **AAR** is the average adhesion rate (the average number of microbes attached to one erythrocyte); **AC** is the coefficient of erythrocytes involved in adhesion (percent of erythrocytes with adhesive microorganisms on their surface).

Concerning the criteria for the evaluation of adhesive properties, a microorganism is considered non-adhesive with IAM $\leq$ 1.75; low-adhesive from 1.76 to 2.5; average-adhesive from 2.51 to 4.0; and high-adhesive with IAM more than 4.0 (Brilis et al., 1986).

The study of the ability of microorganisms to form biofilms was carried out using the plate method by G. O'Toole, 1999. Bacterial cultures were grown in a nutritious broth at a temperature of 37°C. The determination was conducted in flat-bottomed plates for immunoassay analysis. Daily culture of the strains was diluted with a liquid nutrient medium to a concentration of 10^7 CFU/ml, and the resulting suspensions were introduced into 150 µl wells in a plate, except for control wells, receiving medium without microorganisms. Trypcase-soy broth (TSB), produced by bioMerieux, France was used as a liquid nutrient medium. The plate was incubated at 37°C for 24 hours. At the end of the incubation period, the planktonic form of the microorganisms was removed from the wells, the biofilms were washed three times with phosphate-buffered saline buffer (pH 7.2-7.4) in a volume of 200 µl per well, and dried at 37°C for 15 min. Then, 150 μl of 0.1% alcohol solution of crystal violet dye was introduced into each well. The wells filled with dye were incubated at room temperature for 45 minutes. Then the dye that was not adsorbed was removed and the wells were washed three times with distilled water. The wells washed from dye were then filled with 250 μl of 96% ethyl alcohol and left for 45 minutes at room temperature. The optical density of the contents of the wells was measured by StatFax 303 Plus photometer at a wavelength of 630 nm. The intensity of the color of the contents of the wells corresponded to the degree of film formation; while taking into account the optical density of control and test culture. The results were recorded in units of optical density (units). To level the error associated with the optical density of the components of the nutrient medium adsorbed on the plate, the value of the optical density of the control well (sterile nutrient medium) was subtracted from the results obtained for the test samples. Assessment of the data showed that the photosensitizer itself did not have an antimicrobial effect (Table 5.1).

The study of the effect of only LED red radiation showed the following results (Table 5.2).

Table 5.1. Antimicrobial activity of methylene blue photosensitizer at various concentrations with exposure of 30 minutes

Test strains	Photosensitizer concentration, %	Average amount of bacteria, lg CFU/ml	
		Test	Control
P. aeruginosa ATCC 27853	0.1	7.29 ± 0.13	7.33 ± 0.06
P. aeruginosa (n = 9)	0.05	7.32 ± 0.87	7.68 ± 0.57

Table 5.2. Changes in the amount of CFUs of bacteria under the influence of LED red radiation (630 - 650 nm)

Test strains	Average amount of bacteria, lg CFU/ml			
	Control	Test, time of exposure, min.		
		20	40	60
P. aeruginosa ATCC 27853	7.52 ± 0.14	7.51 ± 0.16	7.49 ± 0.15	7.49 ± 0.18
P. aeruginosa (n = 9)	7.27 ± 0.78	7.28 ± 0.13	7.22 ± 0.11	7.21 ± 0.16

Assessment of the effects of LED red radiation demonstrated its insignificant impact on the colony-forming ability of both the reference and clinical isolates of *P aeruginosa.*

Studies on bactericidal activity of photodynamic impact on bacteria *P. aeruginosa* were carried out with local application of methylene blue photosensitizer in the form of 0.1 and 0.05% solution in conjunction with the given parameters of LED radiation: power density – 25 mW/cm^2; time of light exposure – 20, 30 and 40 minutes; energy density – 30 J/cm^2 (light exposure – 20 min) and 60 J/cm^2 (light exposure – 40 min); the area of the irradiated surface comprised about 10 cm^2.

The treatment of microorganisms with photosensitizer followed by LED radiation, that is, PDT effect on microorganisms (30 – 60 J/cm^2), significantly increased bactericidal activity, up to the complete elimination of microbes (Table 5.3). Treatment of test strains with 0.05 - 0.1% solution of methylene blue as early as at energy exposure of 30 J/cm^2 resulted in considerable death of bacteria in the limits of more than 90% ($p < 0.05$). An increase in irradiation exposure up to 30 minutes and energy exposure (45 J/cm^2) for 0.1% PS concentration led to complete elimination of bacteria.

Table 5.3. Changes in the number of CFU test strains of *P. aeruginosa* (n = 10) under the influence of PDT (flux density of radiation energy 25 mW/cm^2)

Concentration without PDT lg CFU/ml	PDT parameters				Concentration after PDT lg CFU
	PS concentration	PS processing time, minutes	Exposure to irradiation, minutes	Energy exposure	
7.12 ± 0.37	0.05%	30	20	30 J/cm^2	5.42 ± 0.27*
7.07 ± 0.29	0.1%	30	20	30 J/cm^2	5.07 ± 0.31*
7.12 ± 0.37	0.05%	30	30	45 J/cm^2	4.21 ± 0.58*
7.07 ± 0.29	0.1%	30	30	45 J/cm^2	0*
7.12 ± 0.37	0.05%	30	40	60 J/cm^2	4.01 ± 0.66*
7.07 ± 0.29	0.1%	30	40	60 J/cm^2	0*

Note: * - the difference is statistically significant ($p < 0.05$).

Thus, the bactericidal effect was not observed in the impact of individual components of the photodynamic therapy complex (photosensitizer methylene blue and irradiation with ultra-bright LEDs in the red region of the optical spectrum) on microbial cells. The treatment of microorganisms with photosensitizer followed by LED radiation, i.e., the effect of PDT on the microorganisms (energy exposure 30-60 J/cm^2), significantly increased bactericidal activity, both at a minimum and at a higher energy density.

The ability of microorganisms to attach to objects, whether they were cells of the epithelium or mucous membranes of the macroorganism, or various abiotic surfaces, is an initial and definitely one of the most important stages of persistence (Usvyatsov, 2006). Therefore, the next stage of the research was the study of the adhesive properties of test strains subjected to sublethal photodynamic effect, namely energy exposure of 30 J/cm^2. The test strains used in the study were characterized by varying adhesion ability: 2 strains of *P. aeruginosa* were low-adhesion; 5, including *P. aeruginosa ATCC 27853* were average-adhesion and 3 were high-adhesion. The results of the study of changes in the adhesion activity of microorganisms under the influence of sublethal photodynamic effects are presented in Table 5.4.

The study showed that in 80% of test strains, the indices characterizing the process of adhesion, under the conditions of sublethal action of PDT, statistically differed from control towards a decrease ($p > 0.05$). Two strains of *P. aeruginosa*, which initially were high-adhesion, after the action of PDT, partially lost this activity and by IAM indicator were classified as average-adhesion.

The next process of adhesion in the life cycle of microorganisms is the process of multiplication and transition to the state of biofilms, well-structured multicellular communities incorporated into the extracellular polymer matrix of its own production. For a large part of the species, this form of existence provides optimal conditions for the implementation of pathogenic and colonization potential, as well as contributes to the preservation of a metabolically inactive part of the population with an extremely low level of susceptibility to antimicrobial agents (Hall-Stoodley L. et al., 2004).

Table 5.4. Indices of adhesion of test strains of *P. aeruginosa* under sublethal photodynamic effects

Test strains according to the adhesion degree	Test conditions	Adhesion indices		
		AAR	AC, (%)	IAM
Low-adhesion (n = 2)	Sublethal effect of PDT	1.57 ± 0.21	68.3 ± 2.67	2.28 ± 0.68
	Control (without PDT)	1.89 ± 0.24	87.6 ± 1.41	2.16 ± 0.42
Average-adhesion (n = 5)	Sublethal effect of PDT	1.46 ± 0.33*	50.0 ± 2.21*	2.91 ± 0.74
	Control (without PDT)	2.49 ± 0.51	87.0 ± 3.12	2.86 ± 0.25
High-adhesion (n = 3)	Sublethal effect of PDT	2.50 ± 0.23*	74.55 ± 3.19*	3.37 ± 0.63*
	Control (without PDT)	4.54 ± 1.16	98.67±4.7	4.61 ± 1.17

Note: * - the difference is statistically significant ($p < 0.05$).

Determination of the biofilm-forming ability of non-irradiated test strains showed that the range of the optical density indices of the extracted dye varied from 0.3915 to 0.9083 units (Table 5.5). Under the influence of the sublethal dose of PDT, the energy exposure of 30 J/cm^2, the specified range narrowed from 0.2992 to 0.7886 units. 80% of strains were shown to have a significant decrease in biofilm activity (Table 5.5).

Thus, the study of persistent properties of test strains subjected to sublethal photodynamic effects showed inhibition of their adhesive properties and biofilm-forming activity.

Based on the experiments conducted *in vitro*, the optimal parameters of the dose of energy density and irradiation exposure were chosen, which determined the choice of application of the components of PDT for the study *in vivo*.

Equally important is the detected property of PDT to influence the strains of bacteria with a high ability to form biofilms, the structure and physiological properties of which provide increased resistance to antimicrobial drugs and help to overcome the protective forces of the macroorganism in infection. Due to the severity of the processes caused by these microorganisms and the almost complete failure of effective strategies for their treatment, new ways to combat active biofilm-forming infectious agents cannot be overestimated.

Table 5.5. Indices of biofilm-forming ability of test strains *P. aeruginosa* under sublethal photodynamic effects

Test strain	Trial conditions	Indices of optical density of the extracted dye, units
1	*2*	*3*
P. aeruginosa ATCC 27853	sublethal effect of PDT	0.4826 ± 0.0021*
	Control (without PDT)	0.7402 ± 0.0009
P. aeruginosa № 8-32	sublethal effect of PDT	0.2992 ± 0.0021*
	Control (without PDT)	0.3915 ± 0.0006
P. aeruginosa № 118	sublethal effect of PDT	0.7886 ± 0.0017
	Control (without PDT)	0.8034 ± 0.0042
P. aeruginosa № 405	sublethal effect of PDT	0.4377 ± 0.0019*
	Control (without PDT)	0.6402 ± 0.0012
P. aeruginosa № 420	sublethal effect of PDT	0.764 ± 0.0013*
	Control (without PDT)	0.9083 ± 0.0016

Note: * - the difference is significant ($p < 0.05$) relative to control.

Debridement of Radiation Ulcers Infected with *Pseudomonas aeruginosa* by PDT (*In Vivo*)

Radiation ulcers are the most severe type of radiation skin and soft tissue lesions. Currently, radiation ulcers develop more often in the late stages after irradiation, and the frequency of their development varies from 5 to 15% (Payne et al., 2008). Treatment of radial ulcers is the most difficult problem with all types of treatment for radiation skin lesions.

Elimination of local infection is a particular problem in the treatment of radiation lesions in general, and especially radiation skin lesions, as it often contaminates skin defects in case of damage to the epidermis: in exudative epidermitis and dermatitis, and especially in radiation ulcers. The control of infection in radiation ulcers is especially important for preventing sepsis. The treatment of such infected skin soft tissues lesions is now considerably complicated by an increasing number of antibiotic-resistant strains of microorganisms (Sharma et al., 2011). One of the most difficult in terms of treatment and complicated course are ulcers, infected with *Pseudomonas aeruginosa* strains.

The purpose of the further stage of the study was to perform a trial to provide evidence of PDT effectiveness in the treatment of simulated radiation skin lesions infected with *Pseudomonas aeruginosa* strains.

Radial skin alteration was modeled on 60 Wistar male rats by local X-ray irradiation of the skin of the right femur at a dose of 85.0 Gy using X-ray therapeutic unit TUR-60. Animal management conditions were in accordance with hygiene requirements and ensured a satisfactory state of health of the subjects. All manipulations were carried out by a competent person in accordance with the requirements of normative documents (Zapadnyuk et al., 1983; The Verkhovna Rada of Ukraine, 2006; The European Parliament and of The Council, 2010). The animals were divided into 2 groups (30 in each group):

- Group A experimental – after the development of radial damage, the surface of the wound was infected with the strain *Pseudomonas aeruginosa* with subsequent PDT;
- Group A control – after the development of radial damage, the surface of the wound was infected with the strain *Pseudomonas aeruginosa*, no therapeutic measures were taken.

Infection of the local radial skin lesions in the animals was performed following the appearance of signs of a radiation ulcer on the 7th day after local irradiation by placing 0.2 ml of the corresponding microbial suspension on the surface of radiation ulcers.

An antibiotic-resistant strain of *P. aeruginosa* №420 highly prone to biofilm formation taken from a trophic ulcer of a patient, who repeatedly received inpatient treatment, was used as a test strain for infection. The antibiotic susceptibility test of the bacterium was conducted using a disc diffusion method and analyzed using the WHONET 5.1 software. The formation of biofilms was investigated by photometric method determining the ability of the microorganism strain to adhesion on the surface of 96-well polystyrene plates, followed by gentian violet staining (O'Toole et al., 1999).

Table 5.6. The effect of photodynamic therapy on the density of microbial colonization in radiation skin ulcer of rats after infection with the reference strain of *P. aeruginosa ATCC 27853*"

Groups of experimental animals	Microbial colonization density (lg CFU/cm^2)					
	Term of observation					
	24 hours after infection	24 hours after PDT	3 days after infection	7 days after PDT	14 days after infection	30 days after PDT
1	2	3	4	5	6	7
Experimental group Radiation ulcer + infection + PDT (n = 9)	5.40 ± 0.24	0	0	0	0	0
1 control group Radiation ulcer + infection (n = 7)	5.66 ± 0.41	4.79 ± 0.25	4.57 ± 0.32	4.43 ± 0.19*	4.26 ± 0.21*	4.11 ± 0.17*
2 control groups Radiation ulcer + infection + methylene blue (n = 7)	5.72 ± 0.32	5.12 ± 0.44	4.82 ± 0.24*	4.67 ± 0.35*	4.12 ± 0.18*	3.85 ± 0.23*
3 control groups Radiation ulcer + infection + photo impact (n = 7)	5.68 ± 0.28	5.41 ± 0.30	5.11 ± 0.26	4.75 ± 0.29*	4.38 ± 0.33*	4.05 ± 0.31*
4 control group Radiation ulcer (n = 10)	0	0	0	0	0	0

Note. * - probably relative to the initial indicator (24 hours after infection).

Table 5.7. The effect of PDT on the time course of *P. aeruginosa* test strain detection on the surface of radiation skin ulcers in laboratory rats

Number of rats in each group, n	Term after infection	Term after treatment (for the experimental group)	Experimental group of animals			Control group of animals		
			Frequency of detection, %	Frequency of complete elimination, %	Population rate, lg CFU/cm²	Frequency of detection, %	Frequency of complete elimination, %	Population rate, lg CFU/cm²
30	1 day	Before PDT	100%	0	5.89 ± 0.3	100%	0	5.77 ± 0.4
30	2 days	1 day after PDT-1	33.3%* **	66.7%	3.15 ± 0.07* **	100%	0	5.58 ± 0.15
30	6 days	5 days after PDT-1	26.6%* **	73.4%	2.28 ± 0.13* **	100%	0	5.48 ± 0.11*
30	7 days	1 days after PDT-2	0* **	100%	0	100%	0	5.37 ± 0.15*
30	13 days	7 days after PDT-2	0* **	100%	0	93.3%	6.7%	4.99 ± 0.21*
24	20 days	14 days after PDT-2	0* **	100%	0	79.2%	20.8%	4.87 ± 0.17*
18	27 days	21 days after PDT-2	0* **	100%	0	66.7%	33.3%	4.32 ± 0.29*
12	36 days	30 days after PDT-2	0* **	100%	0	33.4%	66.6%	3.21 ± 0.25*
6	51 days	45 days after PDT-2	0	100%	0	0	100%	0

Notes: * - the difference is probable relative to the indices before treatment ($p < 0.05$);

** - the difference is probable between the indices of the experimental and control groups ($p < 0.05$).

Preparation of suspensions of microorganisms with a specified concentration of microbial cells was carried out using an electronic device Densi-La-Meter (PLIVA-Lachema a.s., Czech Republic) on the McFarland scale in accordance with the instrument manual and regulatory document (Ministry of Health of Ukraine, 2006). Daily test culture was washed off the surface of agar medium with a sterile physiological solution and adjusted to the number of units of optical density standard according to McFarland, necessary for the trial, for *Pseudomonas aeruginosa* it was 4 units. The number of colony-forming units (CFU) was determined by the method of serial dilutions with subsequent bacterial inoculation on the appropriate nutrient medium.

Samples were collected from the test object using a sterile cotton swab. The delivery time of the material to the laboratory was 10-15 minutes. The material from the swab was thoroughly washed in 1 ml of sterile saline solution and prepared for a series of successive ten-fold dilutions and 0.1 ml of each dilution was inoculated on nutritive agar. The inoculations were cultivated for 18-24 hours at 37°C. Identification of test strains was carried out according to morphological, tinctorial, culture and biochemical properties. The concentration of microorganisms was determined by counting colony-forming units (CFUs) taking into account the dilution of the test material. For convenience, the results of determining the number of microorganisms were expressed in decimal logarithms of the number of microorganisms per cm^2 of the ulcer surface, lg CFU/cm^2.

Photodynamic therapy was carried out using a photonic unit "Barva-LED/630". The power of photon radiation (P) was 25 mW, wavelength (maximum absorption of radiation) was 630 nm. The light exposure time was 30 minutes, energy exposure per session was 45 J/cm^2. Methylene blue was used as a photosensitizer. Wound surface of the radial ulcer was exposed to 0.1% aqueous solution of methylene blue for 30 minutes. The antimicrobial effect of PDT was evaluated based on the results of microbiological research of the material from the surface of skin radial damage over time, namely before therapy, 24 hours after PDT, and after 6, 7, 13, 20, 27, 36 and 51 days after infection. The obtained results were statistically processed by generally accepted methods of statistics using "Statistica v. 8.0" software.

Results of microbiological examination before the treatment showed that the infected radial skin ulcer in the experimental animals had statistically homogeneous indices both in the experimental and in the control groups of the rats. The population level of *Pseudomonas aeruginosa* test strain before the treatment was lg (5.89 ± 0.3) CFU/cm^2 and lg (5.77 ± 0.4) CFU/cm^2 in the experimental and control group, respectively. The data presented in these groups did not differ statistically. PDT treatment was conducted on the second day after infection. The results of the microbiological examination of the surface of the radiation ulcers are presented in Table 5.6. In 24 hours after the treatment test strain of Pseudomonas aeruginosa was found to be completely eliminated. The achieved indices lasted until the end of the trial, namely 30 days after infection. In this case, individual components of antimicrobial PDT (photosensitizer – methylene blue and non-monochromatic radiation in the optical range of 670 nm, 45 J/cm^2 as a light source) did not exert bactericidal impact and did not cause damaging effects. The above-mentioned studies demonstrated the fundamental possibility of high antibacterial activity of PDT according to the developed scheme using one PDT session.

The next task of the study was to increase the pathogenicity of the infectious agent, namely the clinical antibiotic-resistant strain of *P. aeruginosa* № 420 with a high ability to biofilm formation. The results of this trial showed that in the group of rats with radiation ulcers infected with the clinical strain of *P. aeruginosa*, in 24 hours after the PDT session, the colonization of the surface of the ulcer with the test strain was 1.5-2 times less than in the control group. However, the population level of the test strain remained rather high. In the following observation periods, the persistence of bacteria on the surface of the ulcers was maintained, although their number gradually decreased, up to a total elimination on the 21st day. Thus, the aggregate of the received data testified to the insufficiency of only one PDT session and led to an additional session of antimicrobial PDT. In 24 hours after the first PDT session, 66.7% of the experimental animals were not found to have the test strain on the surface of the ulcer ($p < 0.05$). In rats with positive findings of *P. aeruginosa*, the colonization rate of the biotope was lg (3.15 ± 0.07) CFU/cm^2 vs. lg (5.58 ± 0.15) CFU/cm^2 in the control group of animals ($p < 0.05$). In subgroups of animals not undergoing treatment, the population density of the test strains of the bacteria of the examined biotope in 24 hours did not statistically differ from the values at the beginning of the observation period. On the 5th day after the first PDT session, the number of animals in which the test strain was eliminated increased to 73.4% in the group of infected with *P. aeruginosa*. The control animals were found to have the test strain in 100% of cases (Table 5.7).

The second session of PDT in 100% of cases was influenced by the bactericidal effect on the test strain of *P. aeruginosa*. The achieved indices remained until the end of the observation period after the medical measures, namely for 45 days. In the control group of animals, only one case of test strain self-elimination was observed on the 13th day of the trial, and the microbial colonization rates in other rats remained quite high, lg (4.99 ± 0.21) CFU/cm^2 of the ulcer (Table 5.7).

The positive effect of PDT was confirmed by histological, histochemical, and morphometric methods of pathomorphosis of radiation ulcers in rats. The study showed that antibacterial PDT substantially activates cleansing of the wound cavity from necrotic tissues and enhances reparative processes to complete healing of ulcers in experimental rats in contrast to the control group.

CONCLUSION

Microbiological research has experimentally shown that PDT is an effective alternative to antimicrobial therapy in the treatment of infected radiation skin lesions, including cases caused by antibiotic-resistant strains of bacteria with a high ability to biofilm formation. It has been determined that in order to achieve complete elimination of the etiological factor, it is necessary to conduct two sessions of PDT.

REFERENCES

Basnakyan, I. (1992). *Cultivation of Microorganisms with Specified Properties.* Moscow: Medicine, pp. 29-59.

Bjarnsholt, T., Tolker-Nielsen, T., Høiby, N. & Givskov, M. (2010). Interference of *Pseudomonas aeruginosa* Signalling and Biofilm Formation for Infection Control. *Expert Reviews in Molecular Medicine*, Volume *12*, p. e11.

Brilis, V., Brilene, T., Lenzner, Kh. & Lenzner, A. (1986). Methods for Studying the Adhesive Process of Microorganisms. *Laboratory Science*, *4*, pp. 210–212.

Chatzopoulos, G. & Doufexi, A. (2016). Photodynamic Therapy in the Treatment of Aggressive Periodontitis: A Systematic Review. *Medicina Oral, Patología Oral y Cirugía Bucal*, *21*(2), pp. 192-200.

Hall-Stoodley, L., Costerton, J. & Stoodley, P. (2004). Bacterial Biofilms: from the Natural Environment to Infectious Diseases. *Nature Reviews. Microbiology*, *2*(2), pp. 95 – 108.

Hirsch, E. & Tam, V. (2010). Impact of Multidrug-resistant Pseudomonas aeruginosa Infection on Patient Outcomes. *Expert Review of Pharmacoeconomics & Outcomes Research*, *10*(4), pp. 441-451.

Jori, G., Fabris, C., Soncin, M., Ferro, S., Coppellotti, O., Dei, D., Fantetti, L., Chiti, G. & Roncucci, G. (2006). Photodynamic Therapy in the Treatment of Microbial Infections: Basic Principles and Perspective Applications. *Lasers in Surgery and Medicine*, *38*, pp. 468–481.

Kashef, N., Esmaeeli, D., Siroosy, M., Taghi, A., Hesami, F. & Fateh, M. (2011). Photodynamic Inactivation of Drug-resistant Bacteria Isolated from Diabetic Foot Ulcers. *Iranian Journal of Microbiology*, *1* (3), pp. 36-41.

Ministry of Health of Ukraine. (2006). *Information Letter №163-2006 on Innovations in the System of Public Health of Standardization of Preparation of Microbial Suspensions.* Kiev: Ukrmedpatentinform, 10 p.

O'Toole, G., Pratt, L., Watnick, P., Newman, D., Weaver, V. & Kolter, R. (1999). Genetic Approaches to Study of Biofilms. *Methods in Enzymology*, *310*, pp. 91-109.

Payne, W., Naidu, D., Wheeler, C., Barkoe, D., Mentis, M., Salas, R., Smith, D. & Robson, M. (2008). Wound Healing in Patients with Cancer. *Eplasty*, *8*, pp. 45–51.

Plavskiy, V., Plavskaya, L. & Tretyakova, A. (2010). Primary Selection of Photosensitizers for Antimicrobial Photodynamic Therapy from Extracts and Infusions of Medicinal Plants. In: *The 18th International Conference on Laser and Information Technologies in Medicine, Biology and Geoecology*. Novorossiysk: Krasnodar region, p. 24–25.

Pylypenko, M. (2014). Local Radiation Lesions. *Ukrainian Journal of Radiology.*, *22*, (1), pp. 34-45.

Sadykov, R., Bazhenova, L., Kasymova, K. & Sadykov, R. (2011). Possibilities of Methylene Blue in Photodynamic Inactivation of Antibiotic Resistant Strains of Bacteria. *Photobiology and Photomedicine*, *1*, pp. 84–87.

Sharma, S., Dai, T., Kharkwal, G., Huang, Y., Huang, L., De Arce, V., Tegos, G. & Hamblin, M. (2011). Drug discovery of antimicrobial photosensitizers using animal models. *Current Pharmaceutical Design*, *17*, (13), pp. 1303–1319.

Simonova-Pushkar, L., Sklyar, N. & Gertman, V. (2015). Features of Microbiocenosis in Laboratory Animals after Exposure to Ionizing Radiation. *Ukrainian Journal of Radiology*, *23* (1), pp. 23–29.

Stranadko, E. (2011). Photodynamic Therapy of Purulent-inflammatory Diseases of Soft Tissues. *Photobiology and Photomedicine*, *2*, pp. 14–19.

The European Parliament and of The Council. (2010). Directive 2010/63/EU of 22 September 2010 on the Protection of Animals Used for Scientific Purposes. *Official Journal of the European Union*, *276* (10), pp. 33–79.

The Verkhovna Rada of Ukraine. (2006). On Protection of Animals from Cruel Treatment. *Bulletin of the Verkhovna Rada of Ukraine (BVR)*, *27*, p. 230.

Usvyatsov, B., Khustudinova, L. & Pashuta, L. (2006). The Role of Factors of Persistence and Virulence in Microecological Changes in the Human Body. *Journal of Microbiology, Epidemiology and Immunobiology*, *4*, pp. 58-61.

Winkler, K., Simon, C., Finke, M., Bleses, K., Birke, M., Szentmáry, N., Hüttenberger, D., Eppig, T., Stachon, T., Langenbucher, A., Foth, H., Herrmann, M., Seitz, B. & Bischoff, M. (2016). Photodynamic Inactivation of Multidrug-resistant Staphylococcus aureus by Chlorin e6 and Red Light (λ = 670 nm). *Journal of Photochemistry and Photobiology*, *162* (9), pp. 340-347.

Zapadnyuk, P., Zapadnyuk, V. & Zakharia E. (1983). *Laboratory Animals. Breeding, Maintenance, Use in the Experiment*. Kiev: Higher school, pp. 196-223.

In: *Pseudomonas aeruginosa* ISBN: 978-1-53616-100-7
Editors: C. Fah Wong and H. Idris

Chapter 6

Pseudomonas aeruginosa LIPOXYGENASE: A POTENTIAL VIRULENCE FACTOR OF THIS BACTERIUM

Swathi Banthiya, and Hartmut Kuhn*

Charité - Universitätsmedizin Berlin, Corporate Member of Freie Universität Berlin, Humboldt-Universität zu Berlin and Berlin Institute of Health, Institute of Biochemistry, Berlin, Germany.

ABSTRACT

Pseudomonas aeruginosa (*PA*) is one of the most common gram-negative bacteria responsible for a wide variety of life-threatening nosocomial infections in immune compromised patients. Several studies have been dedicated to identify novel virulence factors of this bacterium and a potential candidate is the *Pseudomonas aeruginosa* lipoxygenase (PA-LOX). PA-LOX is a 70 kDa protein that catalyzes the oxygenation of arachidonic acid to 15*S*-hydroperoxy arachidonic acid (15S-HpETE)

* Correponding Author's E-mail: hartmut.kuehn@charite.de.

and linoleic acid to 13S-hydroperoxy linoleic acid (13*S*-HpODE). Its active site is a bifurcated internal cavity and one of the two interconnected subcavities involves the catalytic non-heme iron. PA-LOX is capable of oxidizing membrane phospholipids, which impacts the barrier function of the membranes. Although the biological relevance of PA-LOX is still unclear four major hypotheses have been postulated: i) PA-LOX is involved in the formation of biofilms when the bacteria interact *in vitro* with human lung epithelial cells. ii) Expression of PA-LOX increases the invasive capacity of the bacterium, which might be related to the capability of PA-LOX to oxygenate the membrane lipids of host cells. iii) Recombinant PA-LOX is capable of synthesizing anti-inflammatory lipoxins and this activity might be part of an evasion strategy suppressing the immune system of the host. iv) Since PA-LOX exhibits a low oxygen affinity the enzyme may be involved in oxygen sensing adaptic the metabolic capacity of the bacteria to surrounding oxygen concentrations. This review is aimed at summarizing the catalytic properties of PA-LOX and at critically discussing the different hypotheses for biological relevance of the enzyme.

INTRODUCTION

The opportunistic pathogen *Pseudomonas aeruginosa* (PA), which causes life-threatening infections in immunocompromised individuals, is among the rare bacterial species, which expresses a true lipoxygenase (PA-LOX). This enzyme has first been described more than 50 years ago (Shimahara 1964; Shimahara & Hashizume 1973) but has only recently been comprehensively characterized with respect to its protein-chemical and enzymatic properties (Vance et al. 2004; Vidal-Mas et al. 2005; Lu et al. 2013; Deschamps et al. 2016; Banthiya et al. 2016). In this chapter we will briefly summarize the structural and functional characteristics of PA-LOX and will critically discuss the current hypotheses on the biological relevance of this enzyme.

Pseudomonas aeruginosa (Classification, Properties, Peculiarities)

Pseudomonas aeruginosa (*PA)* is one of the most common gram-negative bacteria and it belongs to the bacterial family of *Pseudomonadaceae* (class: Gammaproteobacteria). It constitutes a rod-shaped aerobic bacterium and is predominantly found in soil and water. However, it can also be detected in plants and animals including humans (Kang et al. 2003). Over the past decades *PA* has emerged as an opportunistic pathogen since it is responsible for a wide variety of nosocomial infections in immune-compromised patients. Typically, such patients suffer from cystic fibrosis, traumatic burns and various sepsis syndromes and because of their compromised immune system they develop on top of their primary disease a *PA* infection. This combination of primary disease and additional *PA* infection may be life threatening and in severe cases the patients die because of multiple organ failure, which is a direct consequence of the *PA* infection (Gellatly & Hancock 2013). Unfortunately, *PA* infections are difficult to treat since this pathogen has developed in the past a high intrinsic resistance to many clinically employed antibiotics. But even when novel antibiotics are employed *PA* has the capacity to quickly develop resistance during the treatment period (Hauser & Ozer 2008).

Modes of Infections

Infections caused by *PA* can be transmitted exogenously from the environment and/or endogenously *via* the gastrointestinal tract (Hauser & Ozer 2008). In hospitals *PA* infections can be spread by jointly consumed contaminated water, by medical devices, by physical contact with infected patients or healthcare workers (Kominos et al. 1972) as well as by inhalation of aerosols (Clifton & Peckham 2010). Another frequent mode of infection is the intake of uncooked fruits and vegetables containing *PA* bound to it on its external surface *via* contaminated soil (Ohara & Itoh

2003). Simple washing of such food before intake may help to prevent infections.

Currently Postulated Virulence Factors of PA

Several studies have been dedicated to identify virulence factors of *PA* that are promising targets for the development of anti-*PA* drugs. The currently known virulence factors of *PA* are divided into two categories (acute or chronic) based on the type of infection they induce (Sharma et al. 2014). Acute virulence factors are either localized on the surface of the bacteria or are secreted into the extracellular space. Surface structures, such as flagella and pili, are essential for cell movement but they also contribute to adherence of the bacteria to the epithelium. The exoenzyme S and other adhesins reinforce this cell adherence. The exotoxin A induces tissue necrosis and phospholipase C is a thermolabile haemolysin, which hydrolyzes membrane phospholipids of host erythrocytes. In addition *PA* expresses at least four different proteases, which have been implicated in tissue necrosis and bleeding (Ben Haj Khalifa et al. 2011).

Other virulence factors are involved in chronic *PA* infections. These proteins include intracellular metabolic enzymes and proteins responsible for the uptake of essential nutrients. For instance, pyoverdin and pyochelin constitute specific siderophores that are needed for bacterial iron uptake and intracellular iron metabolism. More recent data suggested that these siderophores are not only devoted to the iron metabolism of *PA* but that they also display other specific biological functions in the metal homeostasis of this microorganism (Brandel et al. 2012). Some *PA* strains isolated from cystic fibrosis patients produce pseudocapsules of alginate and these structures protect the bacterium from phagocytosis, dehydration and antibiotics. Although the metabolic basis for the formation of these pseudocapsules remains elusive these structures improve adherence to epithelial cells and biofilm formation (Ben Haj Khalifa et al. 2011). The expression of most *PA* virulence factors is controlled by two different types of internal regulatory system: i) the two-component transcriptional regulatory system (Groisman 2001) and ii) the quorum sensing systems (Kariminik et al. 2017). These two mechanisms are essential for survival and proliferation of *PA* in the host.

The formation of biofilms is a very efficient bacterial strategy to survive in the host. In biofilms, bacteria are highly organized and form well-structured communities. For *PA* such communities are frequently encapsulated by excreted polymeric organic material that consists of proteins, lipids, nucleic acids and polysaccharides. These capsules ensure the structural identity of the biofilms and protect the individual bacteria from the immune system of the host (Sharma et al. 2014). The formation of such biofilms on organic matrices (epithelial cells) is characteristic for cystic fibrosis patients (Aanaes et al. 2015) and inhibitors of biofilm formation might function as effective anti-cystic fibrosis drugs. In addition, *PA* also forms biofilms on inorganic matrices, such as on hospital instruments like catheters and endotracheal tubes (inorganic surfaces) (Veesenmeyer et al. 2009). Biofilms on organic and inorganic surfaces continuously keep growing to form mushroom-like networks that facilitate the exchange of nutrients and waste between bacterial cells and make them less resistant towards antibiotics (Kaplan 2010).

LIPOXYGENASES

Lipoxygenases (LOXs) form a family of fatty acid dioxygenases, which catalyzes the oxygenation of polyunsaturated fatty acids containing one or more bisallylic methylene(s) to their hydroperoxy derivatives (Haeggström & Funk 2011; Wasternack & Feussner 2018). These hydroperoxy fatty acids are further metabolized to secondary and tertiary products and these downstream products have extensive roles in diverse aspects of plant physiology such as pathogen resistance and organ development (Porta & Rocha-Sosa 2002). In mammals, LOXs have been implicated in the biosynthesis of pro-inflammatory eicosanoids (Haeggström & Funk 2011) but also in the formation of anti-inflammatory lipoxins (Romano 2010), resolvins (Yoo et al. 2013), maresins (Serhan et al. 2012). The catalytic center of most LOXs contains a non-heme iron that is essential for its catalytic activity. However, some fungal LOXs involve a manganese ion instead. For instance, the LOX from *Gäumannomyces graminis* contains 1 manganese ion per molecule enzyme (Su & Oliw 1998). For iron containing LOXs, the metal ion shuttles between its ferrous (Fe^{2+}) and its ferric (Fe^{3+}) form during the catalytic cycle. When expressed in cells iron containing LOXs are present as inactive ferrous enzymes and need to be oxidized to the ferric (Fe^{3+}) form (Pistorius & Axelrod 1974; Funk et al. 1990) to initiate catalysis. The oxidation of the iron is accomplished by organic and inorganic peroxides, which are formed during cellular metabolism. Thus, the cellular redox state constitutes an important regulator of the intracellular LOX pathway.

The human genome involves six functional LOX genes (*ALOX15, ALOX15B, ALOX12, ALOX12B, ALOXE3, ALOX5*) that encode for six distinct LOX isoforms. The mouse genome involves single orthologs for all human LOX genes. However, in this species the *Aloxe12* gene, which is a corrupted pseudogene in humans, is functional and encodes for a additional LOX-isoform (Funk et al. 2002). For a long time, the presence of LOXs was considered a eukaryotic affair since such enzymes have mainly be detected in mammals, plants, lower marine invertebrates and fungi (Horn et al. 2015). However, more recently, LOX sequences have also been detected in various prokaryotic organisms, but their biological relevance is still unknown (Hansen et al. 2013).

Bacterial Lipoxygenases and Their Frequency of Occurrence

So far LOXs have been identified in two (eukaryotes, bacteria) of the three domains of life on earth. No functional LOX genes have been identified in archaea and viruses (Horn et al. 2015). A systematic search of the ~13,000 bacterial genomes sequenced so far suggested the presence of LOX-like sequences in some 60 bacterial species indicating that LOXs are not widely distributed in bacteria. In fact, all major human pathogenic bacteria lack functional LOXs genes. Such low prevalence (<0.5%)

indicates that LOXs might not be essential for bacterial survival and for human pathology (Horn et al. 2015). From the evolutionary point of view, it is worth mentioning that there has not been a systematic evolution of LOXs in prokaryotes (Hansen et al. 2013). In fact, bacteria have probably acquired LOX sequences from eukaryotes over the years *via* horizontal gene transfer and some bacteria have retained them as functional genes (Porta & Rocha-Sosa 2001). Other bacteria have abundant these genes since these sequences did not confer to them an evolutionary advantage. For those bacteria carrying functional LOX genes there appears to be an evolutionary pressure on these genes, which prevented inactivating mutations and subsequent destruction of the genes. The bacterial species carrying apparently functional LOX genes include firmicutes, different types of actinobacteria, cyanobacteria, proteobacteria and representatives of CFB group (Horn et al., 2015). In *PA* several LOX alleles have been detected (Hansen et al. 2013) and the corresponding enzymes might contribute to the pathogenicity of *PA* (Vance et al. 2004).

Discovery and Partial Characterization of *Pseudomonas aeruginosa* Lipoxygenase

For many years, researchers have assumed that bacteria lack LOXs because *E. coli*, which constitutes a bacterial model organism, does not involve LOX-like sequences in its genome. On the other hand, already in 1964 a lipid oxidizing enzyme was identified in *PA* (Shimahara 1964; Shimahara & Hashizume 1973) but it remained unclear whether this protein was a true LOX. Lower organisms, such as bacteria, fungi, mosses and others express different kinds of lipid oxidizing enzymes and from the oxygenation products it is not always straightforward to conclude, which enzymes may have been involved. For instance, the fungus *Aspergillus nidulans* expresses a fusion protein, which oxidizes lipids according to a completely different reaction mechanism (Brodhun et al. 2009). In 1997, Guerrero et al. reported the oxygenation of oleic acid to (E)-10-hydroxy-8-octadecenoic acid, (E)-10-hydroperoxy-8- octadecenoic acid and (E)-7,10-

dihydroxy-8-octadecenoic acid (Guerrero et al. 1997) but here again the oxidation mechanism remained unclear. This observation was later on confirmed by Busquets et al. who reported that the oleic acid oxidizing enzyme exhibited a molecular weight of 45 kDa (Busquets et al. 2004). At that time and even today it has not really been clarified whether this periplasmic oleic acid oxidizing enzyme is a variant of the PA-LOX (Vance et al. 2004).

Although the first reports on the presence of a lipid oxidizing enzyme in *PA* appeared already in the mid 1960s and 1970s (Shimahara 1964; Shimahara & Hashizume 1973) the first convincing evidence for the expression of a true LOX in this bacterium was published in 2004 when Vance and co-workers identified a LOX gene in the genome of *PA*. They expressed and characterized a gene technically modified version of the enzyme in *E. coli* (Vance et al. 2004). This LOX exhibited a molecular weight of 70 kDa and oxygenated arachidonic acid and linoleic acid to 15*S*-H(p)ETE and 13*S*-H(p)ODE, respectively. Later on other researchers expressed similar enzyme variants in *E. coli* and characterized the enzymes with respect to its protein-chemical and enzymatic properties (Vance et al. 2004; Vidal-Mas et al. 2005; Lu et al. 2013; Deschamps et al. 2016; Banthiya et al. 2016). In 2013 the 3D-structure of PA-LOX was solved at a molecular resolution of 1.75 Å (Garetta et al., 2013) and the authors reported that the active site of the enzyme was occupied by an endogenous phospholipid ligand, which was co-purified with the recombinant enzyme. In 2016, Deschamps (Deschamps et al. 2016) and Banthiya (Banthiya et al. 2016) expressed PA-LOX as a recombinant His-tag fusion protein in *E.coli* and characterized the enzyme in great detail.

Protein-Chemical Properties of PA-LOX

Enzymes are catalytically active proteins, which accelerate chemical reactions in living systems. Like non-enzymatic catalysts enzymes increase the rate of a chemical reaction by lowering the activation energy and by decreasing the time period required for adjustment of the chemical equilibrium. Impacting these two parameters some enzymes can make the substrate-to-product conversion several million times more efficient. In this section, we will compare the protein chemical properties of PA-LOX with the corresponding data of eukaryotic enzymes.

Amino Acid Alignments with Other LOX

In order to explore the degree of similarity PA-LOX shares with eukaryotic LOX-isoforms, we conducted dual amino acid sequence alignments with selected plant and mammalian ALOX-isoforms. For this purpose, the amino acid sequence of PA-LOX (without it's secretory sequence) was aligned with selected eukaryotic LOX-isoforms (mouse Alox15, soybean LOX1, *A. thaliana* LOX2, rabbit ALOX15, human ALOX15 and human ALOX15B). The results shown in Table 6.1 indicate that this bacterial protein shares a low degree of amino acid similarity with all of these LOX-isoforms. Moreover, these data show that PA-LOX may not be closely related to these 12/15-lipoxygenating LOX-isoforms of vascular plant and mammals. A low degree of amino acid conservation was also observed when PA-LOX was compared with LOX2 of *A. thaliana*. This finding was somewhat surprising since the *A. thaliana* enzyme represents the family of chloroplastic LOXs and since chloroplasts and bacteria share a common ancestor.

Table 6.1. Amino acid sequence alignment of PA-LOX with eukaryotic LOX isoforms

LOX isoform	Molecular weight (kDa)	Accession number	% Identity	% Similarity
Mouse Alox15	75.45	CAJ18471.1	25	39
Arabidopsis thaliana LOX2	85.90	OAP05844	21	31
Rabbit ALOX15	75.26	NP_001075751	22	35
Human ALOX15	74.67	NP_001131.3	26	39
Human ALOX15B	75.86	NP_001132.2	28	41

Recombinant Expression and Comprehensive Characterization

PA-LOX is the only LOX reported so far to be a secretory protein in its native form. As indicated in Figure 6.1, the 19 N-terminal amino acids of this enzyme may function as secretion signal peptide as suggested by *in silico* prediction algorithms (Vance et al. 2004). However, convincing experimental confirmation of the functionality of this hypothetical secretion sequence is still pending.

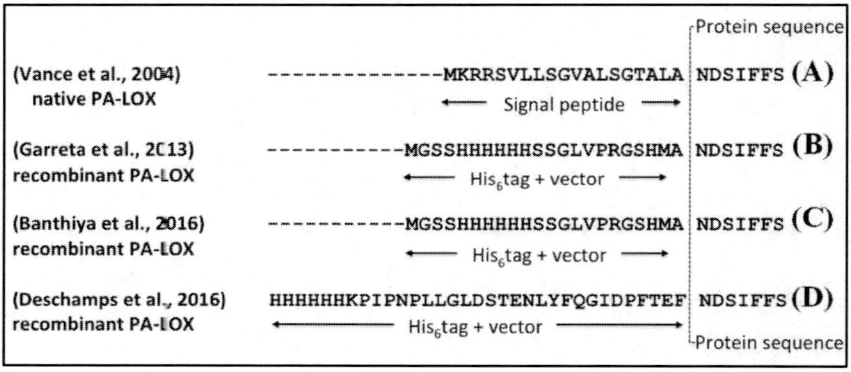

Figure 6.1. The predicted signal peptide of native PA-LOX.

PA-LOX was expressed as recombinant protein in several studies with (Vance et al. 2004; Lu et al. 2013) and without (Garreta et al. 2013; Deschamps et al. 2016; Banthiya et al. 2016) its signal peptide. For technical reasons, we omitted the putative signal sequence in our studies and replaced it with an N-terminal hexa-His-tag followed by 11 additional amino acids of the prokaryotic expression plasmid pET 28 (Figure 6.1C). After purifying the enzyme from the bacterial lysate supernatant by affinity chromatography, SDS-PAGE indicated >80% pure protein in the elution fractions 1-3. Quantification of the total yield of the expression/purification system revealed that 63.3 ± 5 mg of pure PA-LOX were obtained from 1 liter (3.2 ± 0.3 mg PA-LOX/50 ml culture volume) of bacterial culture fluid (Banthiya et al. 2016). In another study (Deschamps et al. 2016) a similar expression construct lacking the N-terminal signal peptide was created. Here the signal peptide was replaced by an N-terminal hexa-his-tag followed by 27 additional amino acids of the pET151/D-TOPO vector (Figure 6.1D). This enzyme construct was expressed in *E. coli* using a different expression system and the authors reported a total yield of 12 mg purified PA-LOX per liter of bacterial culture fluid (Deschamps et al. 2016). Although the expression/purification yields were different in the two studies, the recombinant enzymes were characterized as 15-S-lipoxygenating enzymes. Linoleic acid (LA) was converted by the two enzyme preparations to 13*S*-HODE (Banthiya et al. 2016; Deschamps et al. 2016; Lu et al. 2013; Vance et al. 2004). Similar reaction products were also identified for the *Nostoc punctiforme* ATCC 29133 LOX-isoforms NpLOX1 and NpLOX2 (Koeduka et al. 2007) and these data have been confirmed by Lang et al. (Lang & Feussner 2007). In contrast, the LOX from *Nostoc sp.* PCC 7120 catalyzed oxidation of linoleic acid to 9*R*-HODE (Lang et al. 2008; Zheng et al. 2008).

A systematic search for LOX-like sequences in the genomes of different *Myxococcus* species indicated the presence of three potential LOX genes in *M. xanthus* (WP_011551853.1, WP_011551854.1, SEU34910.1). The protein encoded for by the WP_011551853.1 gene was identified experimentally as functional acidic LOX-isoform (Qian et al., 2017). Recombinant expression of the WP_011551854.1 gene also yielded a catalytically active enzyme (An et al., 2018), which exhibited distinct catalytic properties. When we search the genome of another *Myxococcus* species (*M. fulvus)* we found that the SEU34910.1 gene, which was previously identified in *M. xanthus*, was also present in *M fulvus*. In addition, we detected another putative LOX gene (WP_046712474.1), which has not been characterized so far. To explore whether the WP_046712474.1 gene encodes for a functional LOX, we expressed the corresponding enzyme in different pro- and eukaryotic expression systems as N-terminal his-tag fusion protein and characterized the recombinant enzyme with respect to its protein-chemical and enzymatic properties. Our results (Goloshchapova et al., unpublished data) indicated that *M. fulvus* expresses an arachidonic acid 12*S*-lipoxygenating LOX-isoform (MF-LOX1), which only shares a low degree of catalytic similarity with the recently characterized enzymes from *M. xanthus* (An et al., 2018, Qian et al., 2017).

Molecular Weight of PA-LOX

The theoretical MW (tMW) of our PA-LOX preparation, which was calculated on the basis of the amino acid composition, was 74.98 kDa. This value is similar to the tMW of native PA-LOX involving the putative secretory sequence (74.53 kDa). The experimental molecular weight (MW) of PA-LOX obtained from our expression system was 70 kDa (Banthiya et al. 2016) and these data are consistent with previous results (Lu et al. 2013). However, in the Deschamps study (Deschamps et al. 2016) the enzyme preparation exhibited an experimental MW of 76 kDa. Table 6.2 summarizes the MWs of the PA-LOX preparations obtained in different studies. It remains unclear why the Deschamps construct showed a MW, which is about 6 kDa higher than that of our preparation. The 12 additional amino acids (1320 kDa) in this expression construct are not sufficient to explain the difference. However, the vector sequence of the "Deschamps enzyme" has four proline residues while our construct carries just one proline. Prolines are known to compromise the flexibility of proteins as they introduce natural kinks into the polypeptide chains. These kinks might lower the electrophoretic mobility of the protein, which can be misinterpreted as higher MW (Kirkland et al. 1998).

Table 6.2. Molecular weights of different PA fatty acid oxidases reported in different studies

Study	Molecular weight (kDa)	
	Experimental (eMW)	Theoretical (tMW)
(Busquets et al. 2004)	45	no sequence
(Lu et al. 2013)	70	74.80
(Garetta et al., 2013).	70	74.98
(Deschamps et al. 2016)	76	76.62
(Banthiya et al. 2016)	70	74.98

When Busquets and co-workers isolated the oleic acid oxidase from the periplasmic space of the PA strain 42A2 they determined an eMW of about 45 kDa (Busquets et al. 2004). Unfortunately, no sequence information has been provided for this enzyme and thus, it remains unclear whether this protein (45 kDa) is somehow related to PA-LOX (75 kDa). If PA-LOX is the precursor protein of the oleic acid oxidase it must have been cleaved by proteases before or after secretion into the periplasmic space. For PA-LOX such proteolytic cleavage is only possible in the N-terminal region of the protein as proteolytic digestion of C-terminal sequences would results in the loss of iron ligands. Such cleavage would completely inactivate the enzyme. While the eMW of most recombinant PA-LOX preparations (70-76 kDa) is comparable with the corresponding values of mammalian LOXs (Brash 1999), other bacterial species express enzymes, which are considerably bigger (Iny et al. 1993).

Isoelectric Point

When we calculated theoretical the isoelectric point (pI) of our recombinant PA-LOX preparation on the basis of the primary structure of the enzyme using the Expasy online tool we obtained a pI of 6.18. For this purpose we carried out isoelectric focusing with our electrophoretically pure PA-LOX preparation and observed that the homogenous band we detected in SDS-PAGE was split into three distinct bands migrating in the pH range 5.5 - 6.2 (Figure 6.2). These data indicate a structural microheterogeneity of our enzyme preparation. Corresponding analysis with native rabbit ALOX15 revealed a single protein band, which migrated with a pI of 5.5 (data not shown) and this data excludes methodological artifacts. The molecular basis for the structural microheterogeneity has not been explored but it might be possible that during recombinant expression the polypeptide chain folds into slightly different 3D structure, which is characterized by different pI values. On the other hand, different types post-translational modifications, which hardly impact the molecular weight, might contribute to the microheterogeneity that was observed in isoelectric focusing.

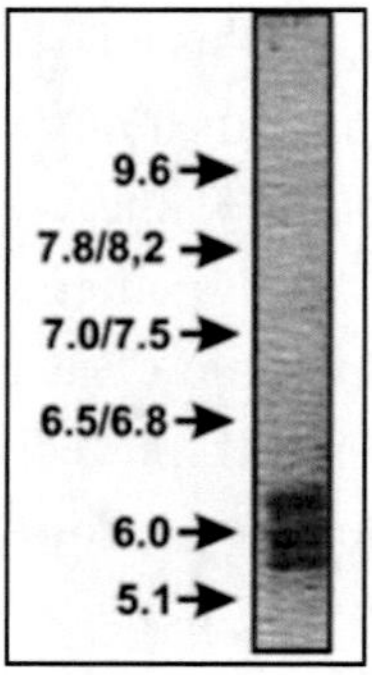

Figure 6.2. Isoelectric focusing (IF) of PA-LOX. In order to determine the native isoelectric point of PA-LOX, we carried out its IF. For this purpose a pre-casted IF gel (Bio-Rad, Munich, Germany) was employed and IF was run for 2.5 h on a Bromma LKB 2197 High Voltage Power Supply electrophoresis system. Protein bands were stained with Coomassie blue and the IF standards employed for this study were the following: phycocyanin, IP 4,45-4,75; β-lactoglobulin B, IP 5,1; bovine carbonic anhydrase, IP 6,0; human carbonic anhydrase, IP 6,5; equine myoglobin, IP 6.8/7.0; human hemoglobin A, IP 7.1; human hemoglobin C IP 7.5; lentil lectin IP 7.8/8.0/8.2 and cytochrome c IP 9.6.

Iron Content of PA-LOX

Most LOXs characterized so far contain a non-heme iron as catalytically active transition metal at their active sites. However, some fungal LOXs involve a manganese ion instead. For instance, LOX from *Gäumannomyces graminis* contains 1 manganese ion per molecule enzyme (Su & Oliw 1998) and the manganese ion also undergoes valency shuttling during the catalytic cycle. The iron content of PA-LOX obtained in our study was 106 ± 3 mol % (Banthiya et al. 2016). This data indicates that each PA-LOX molecule in our preparation contained one iron ion. There was no need for external iron supplementation when expressing the recombinant enzyme. In the Deschamps study an iron load of the final enzyme preparation of 67% was calculated (Deschamps et al. 2016). The 45 kDa PA-LOX isolated from the periplasmic space showed an iron load of 55%. In contrast, the manganese concentration in this enzyme preparation was much lower, so that a manganese load of only 2.5% was calculated (Busquets et al. 2004).

Secretion Sequence (Subcellular Distribution)

In order to explore if native PA-LOX is a secretory protein, the wildtype PA-LOX sequence (accession number: AIC73832.2) was analyzed with the online prediction tool SignalP (Nielsen 2017). This software predicts the presence and location of signal peptide cleavage sites in amino acid sequences of proteins of many organisms.

Upon submission of the PA-LOX sequence, based on scores received, the prediction tool identified two interesting peculiarities: i) There is a proteolytic cleavage site between amino acid Glu21 and Ser22. ii) The amino acids between Met1 and Asn20 form a putative signal peptide as suggested by Figure 6.1A. Although the online prediction tool allows fairly reliable conclusions it should be stressed that the functionality of the secretion sequence of PA-LOX has not been confirmed experimentally.

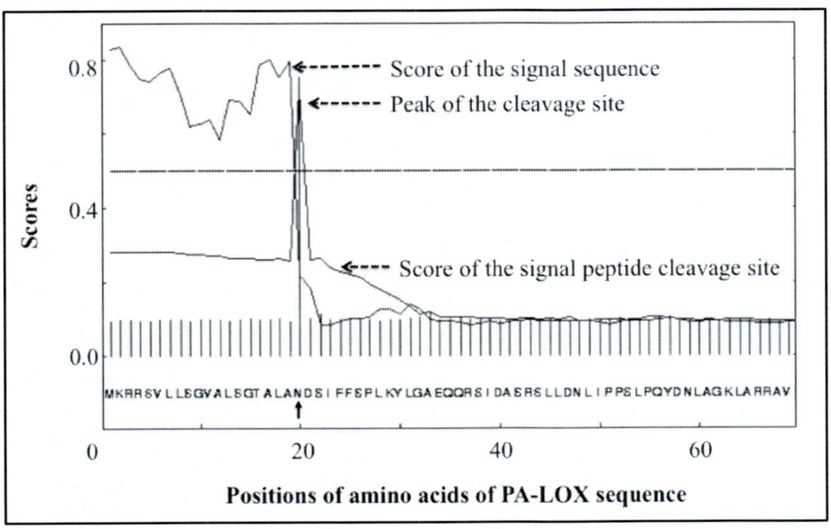

Figure 6.3. Putative signal sequence of PA-LOX predicted by SignalP.

In another study the N-terminal 19 amino acids were suggested to form a signal peptide (Vance et al. 2004) but here again, the sequence-based prediction has not been tested experimentally. There are several ways to do so but the most direct approach would be to fuse the cDNA of the putative signal sequence upstream to a reporter protein (e.g., green fluorescence protein) and then to follow localization of the fluorescence after expression of this reporter construct.

ENZYMATIC PROPERTIES OF PA-LOX

The functionality of all enzymes, in particular the catalytic efficiency, is highly sensitive to change in the conditions (like pH, temperature, structure of substrate) of the surrounding medium. In this section we focus on discussing the various enzymatic properties of PA-LOX and compare these data with the corresponding parameters determined previously for eukaryotic lipoxygenases.

Thermostability and Activation Energy of PA-LOX

When kept at 4°C, recombinant PA-LOX looses 60% of its catalytic activity within the first 24 hours of storage irrespective of the presence or absence of glycerol as stabilizing agent. After one week of storage at 4°C, it exhibited 20% of its initial activity as residual activity (Banthiya et al. 2016). In contrast, the 45 kDa PA-LOX isolated from the periplasmic space of *PA* appeared to be more stable as it exhibited 100% of its initial activity even after one week of storage at 4°C (Busquets et al. 2004).

The temperature profile of PA-LOX catalyzed fatty acid oxygenation has been determined in several studies. The temperature optimum (T_{opt}) of PA-LOX catalyzed linoleic acid oxygenation is 25°C (Lu et al. 2013; Deschamps et al. 2016). A similar T_{opt} was reported for the PA-LOX isolated from periplasmic space of the bacterium (Busquets et al. 2004). The PA-LOX preparation obtained in our study was less temperature sensitive and exhibited a higher T_{opt} of 35°C. From the Arrhenius plot of the catalytic activities in the temperature range of 5 - 35°C an activation energy of 33.8 kJ/mol x K was determined. This value is more than twice as high as the corresponding activation energy determined for rabbit ALOX15 (14.8 kJ/mol x K.), which was quantified in side-by-side experiments (Banthiya et al. 2016).

Substrate Specificity and Product Profile of PA-LOX with AA and LA

PA-LOX oxygenates both arachidonic acid (AA) and linoleic acid (LA) at the n-6th position of the fatty acid backbone. It catalyses the conversion of AA to 15*S*-H(p)ETE and of LA to 13*S*-H(p)ODE, thus qualifying the enzyme as n-6 *S*-LOX (Vance et al. 2004). A similar product profile (Koeduka et al. 2007) was also observed for the two LOX isoforms identified in the cyanobacteria *Nostoc punctiforme* ATCC 29133 (NpLOX1 and NpLOX2). In contrast, the LOX isoform from *Nostoc sp.* PCC 7120 catalyzed oxygenation of LA to 9*R*-HODE (Lang et al. 2008). Spectrophotometric analysis of the rate of AA and LA oxygenation by PA-LOX indicated that PA-LOX exhibits a higher turnover number with LA than with AA (65.4 s^{-1} for LA *vs.* 23.5 s^{-1} for AA). On the other hand, PA-LOX exhibits a lower affinity for LA binding when compared with AA (Km of 174 for LA vs. 48.9 μM for AA). If one combines these two basic kinetic characteristics it becomes evident that PA-LOX exhibits a similar catalytic efficiency (0.38 for LA vs. 0.48 $(\mu M \times s)^{-1}$ for AA) with these two polyenoic fatty acids (Banthiya et al. 2016). The catalytic efficiency of PA-LOX reported in the Deschamps study [16 $(\mu M \times s)^{-1}$] was significantly higher than that determined in our experiments (Deschamps et al. 2016). The reason for this difference might be related to the different assay conditions employed in the two studies. In the Deschamps study activity assays were carried out at lower pH (6.5 *vs.* 7.4), at lower substrate concentrations (10 μM *vs.* 100 μM) and in the presence/absence of detergents (0.01% Triton X-100 vs. no detergent) (Deschamps et al. 2016).

In the absence of detergents a similarly low catalytic efficiency of 2 $(\mu M \times s)^{-1}$ was determined for human ALOX15 (Wecksler et al. 2008).

Oxygenation of Other Polyunsaturated Fatty Acids (AA, LA, Lna, DHA, EPA)

In mammalian cells arachidonic acid (AA), linoleic acid (LA), alpha- and gamma-linolenic acid (LnA), eicosapentaenoic acid (EPA) and docosahexaenoic acid (DHA) are among the most abundant naturally occurring LOX substrates and these fatty acids are widely distributed in mammalian cell membranes. In contrast, polyenoic fatty acids are rare in bacteria.

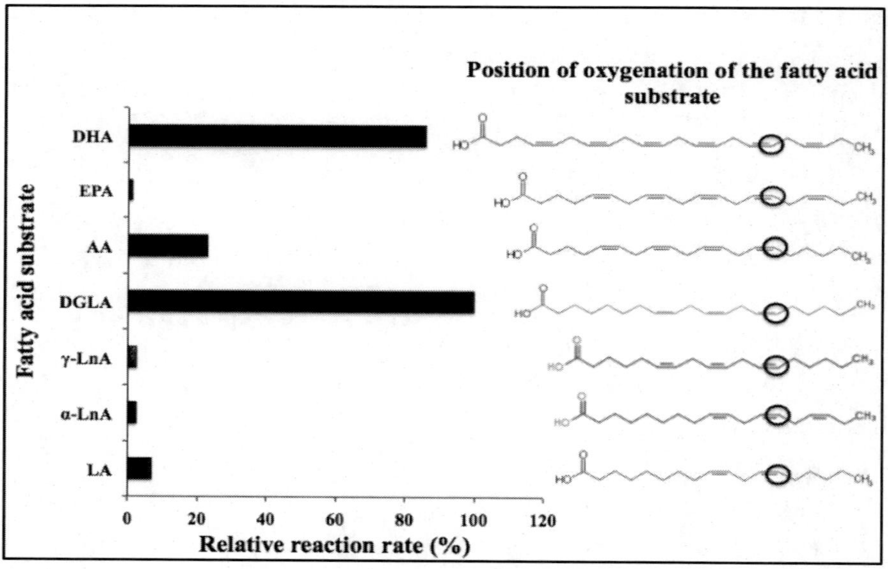

Figure 6.4. Relative catalytic activities of PA-LOX with common mammalian polyenoic fatty acid substrates. The black circles indicate the positions of oxygen insertion into the fatty acid backbone; AA- Arachidonic acid, DGLA-Dihomo-γ-Linolenic acid, DHA-Docosahexaenoic acid, EPA- Eicosapentaenoic acid, LA-Linoleic acid, α-LnA – α -Linolenic acid, γ-LnA - γ-Linolenic acid).

Spectrophotometric assays measuring the increase in absorbance at 235 nm with PA-LOX and the above mentioned substrates followed by GC/MS analysis of the reaction products indicated that PA-LOX oxygenated all the above substrates at the n-6th carbon atom of the fatty acid backbone irrespective of the differences in chain length and the number of double bonds. As indicated by the relative reaction rates (Figure 6.4) the order of preference of the fatty acid substrates was as follows: DGLA > DHA > AA > LA > EPA (Banthiya et al. 2016). While EPA and LA are well oxygenated by many mammalian ALOX isoforms (Haeggström & Funk 2011) our PA-LOX preparation does hardly accept these substrates. However, under different assay conditions the order of preference was AA > LA > EPA > DHA > DGLA (Deschamps et al. 2016). These differences (Deschamps et al. 2016; Banthiya et al. 2016) are likely to be related to the different assay conditions used in the two studies (different substrate concentrations: 10 μM in the Deschamps study *vs.* 100 μM in our study; different pH values: 7.0 in the Deschamps study *vs.* 7.4 in our study; presence and absence of detergents).

Recently, Kutzner et al. conducted an interesting study, in which they tested the substrate selectivity of several mammalian 12- and 15-lipoxygenating ALOX15 orthologs (human, chimpanzee, orangutan, rabbit, pigs mouse, rat and rhesus monkeys). For their experiments, the authors incubated the recombinant LOX preparations with equimolar mixtures of AA, EPA and DHA. Quantification of the products indicated that most mammalian ALOX15 orthologs prefer DHA >EPA > AA. However, the ALOX15 orthologs of mice and rhesus monkeys selected EPA as best substrate (Kutzner et al. 2017). These results are somewhat different from the observations seen with the two PA-LOX preparations reported above. While the PA-LOX preparation from the Banthiya study also preferred DHA over AA and EPA (Banthiya et al. 2016), inverse observations were made with the Deschamps enzyme (Deschamps et al. 2016). The reasons for such differences in PUFA preference of the two enzyme preparations have not been explored but here again the different assay concentrations might contribute.

Phospholipids as Substrate

ALOX15 orthologs of different mammalian species are capable of oxidizing complex ester lipids carrying polyunsaturated fatty acids even if these lipids are incorporated in complex lipid-protein assemblies (Schewe et al. 1975; Belkner et al. 1991; Takahashi et al. 1993; Pekárová et al. 2015). Unfortunately, for the time being it remains unclear why rabbit ALOX15 exhibits such pronounced phospholipid oxygenase activity. The putative substrate-binding pocket of this enzyme is a single hydrophobic cavity that is connected to the protein surface and involves the catalytic non-heme iron. *In silico* substrate docking studies suggested that this cavity can easily adopt a fatty acid molecule but there is not enough space to bind a complete phospholipid without major rearrangement of the protein. Moreover, there is no separate subcavity that can adopt the polar head group of such substrates. Thus, the mode of phospholipid binding at the active site of rabbit ALOX15 needs to be explored in more detail.

In contrast, the crystal structure of PA-LOX indicated (see chapter 4.1) a bifurcated substrate-binding pocket that can easily adopt a phospholipid molecule. These data suggested the principle capability of this enzyme to bind and oxygenate phospholipids. To quantify the phospholipid oxygenase activity of PA-LOX, we incubated the recombinant enzyme *in vitro* with different preparations of naturally occurring phospholipids (phosphotidyethanolamine, phosphatidylcholine, phosphatidylserine and phosphatidylionositol). After the incubation period, we hydrolyzed the phospholipids and analyzed the oxygenated fatty acid derivatives by HPLC. As positive control rabbit ALOX15 was tested in site-by-site. In the two incubations (PA-LOX, rabbit ALOX15) we detected significant amounts of 13-HODE and 15-HETE by RP-HPLC and chiral phase HPLC indicated a strong preponderance of the *S*-isomer. Taken together these results indicated that PA-LOX exhibits a phospholipid oxygenase activity but the catalytic efficiency of rabbit ALOX15 was 10-fold higher than that of the PA enzyme. Moreover, it should be stressed that the fatty acid oxygenase activity of the two enzymes was much higher than their phospholipid oxygenase activities (Banthiya et al. 2016).

The crystal structure of PA-LOX indicated that its substrate-binding pocket is a bifurcated cavity, which is connected to the protein surface. However, the entrance into this cavity is blocked by two long helices, which may function as lid structure. In other words, the substrate-binding cavity of this enzyme consists of two separated sub-cavities that are interconnected by a joining "lobby". Sub-cavity 1 of this pocket involves the non-heme iron and the sn1 fatty acid of the endogenous ligand. Sub-cavity 2 contains the sn2 fatty acid of the endogenous ligand, which cannot be oxygenated owing to the absence of iron in this sub-cavity. Since most mammalian unsaturated phospholipids usually contain their polyenoic fatty acid at the sn2 position, it might be possible that during a reaction with phospholipids, the unsaturated fatty acid is bound in the wrong (iron deficient) sub-cavity. Such unproductive substrate binding might lower the reaction rate of phospholipid oxygenation. If this hypothesis is true, a synthetic phospholipid carrying linoleic at the sn1 position should be a better substrate for PA-LOX than a phospholipid molecule involving linoleic acid at the sn2 position. Unfortunately, phospholipids carrying the linoleic acid at the sn1 position are not commercially available and chemical synthesis is quite laborious and expensive. Thus, corresponding comparative experiments cannot be carried out at the moment. However, to explore whether PA-LOX prefers oxygenation of the sn1 or the sn2 fatty acids, when both positions are occupied by a linoleic acid residue, we next compared 1-palmitoyl-2-linoleoyl-phosphatidylethanolamine and 1,2-dilinoleoyl-phosphatidylethanolamine as substrates for PA-LOX and found that the reaction rate of 1,2-dilinoleoyl- phosphatidylethanolamine oxygenation was 7.2 ± 1.0-fold higher than that of 1-palmitoyl-2-linoleoyl-phosphatidylethanolamine. HPLC analysis of the reaction products indicated that 1,2-dilinoleoyl-phosphatidylethanolamine was oxygenated at the sn1 and sn2 position forming 1,2-di-HpODE- phosphatidyl-ethanolamine. Taken together, these data indicated that PA-LOX is capable of oxygenating phospholipids at their sn1 and their sn2 position.

Biomembranes as Substrates for PA-LOX

Phospholipids are essential constituents of biomembranes and the capability of PA-LOX to oxygenate isolated phospholipids raised the question whether this enzyme is also capable of oxygenating biomembranes. In order to quantify the membrane oxygenase activity of PA-LOX, the recombinant enzyme was incubated *in vitro* with model membrane preparations (vesicles of mitochondrial inner membrane). Afterwards, lipids were extracted, hydrolyzed and the oxygenated free fatty acid derivatives were analyzed by RP-HPLC (Banthiya et al. 2015). These analyses indicated the formation of esterified 13*S*-H(p)ODE and 15*S*-H(p)ETE (oxygenated products of linoleic acid and arachidonic acid, respectively), which were absent in the control mixtures lacking the PA-LOX enzyme. Side by side incubations with native purified rabbit ALOX15 indicated that the membrane oxygenase activity of the rabbit enzyme was 4-fold higher than that of PA-LOX when similar arachidonic acid oxygenase activity were adjusted (Banthiya et al. 2015).

The Lipoxin Synthase Activity of PA-LOX

Lipoxins are secondary products of arachidonic acid oxygenation that function as important immune modulators (Romano 2010). Previous studies have shown the ability of rabbit ALOX15 to synthesize lipoxins by catalyzing the oxygenation 15S-HETE methyl ester and of 5,15-DiHETE to lipoxin isomers (Kuhn et al. 1987). To explore whether PA-LOX is also capable of producing lipoxins we incubated recombinant PA-LOX with 5*S*-HETE and with a 1:1 mixture of 5*S*,6*S*- and 5*S*,6*R*-DiHETE and analyzed the product mixture by RP-HPLC recording the absorbance at 300 nm. Here we found that both substrates were converted to lipoxin A and lipoxin B isomers (Banthiya et al. 2016). In contrast, in the Deschamps study (Deschamps et al. 2016) the authors reported that PA-LOX does not exhibit a lipoxin synthase activity. The reasons for the controversial results will be discussed later on in this review.

STRUCTURAL PROPERTIES OF PA-LOX

To effectively catalyze a chemical reaction, substrate(s) are bound at the active site of enzymes. This substrate binding is highly specific and involves in most cases non-covalent enzyme-substrate interactions. The interactions arrest the substrate in a certain conformation at the active site, which energetically favors the formation of the catalytic transition state. The active site of most enzymes has a defined three-dimensional structure, which forces the substrate to adopt a certain conformation. In this section we will review the structural peculiarities of PA-LOX and will describe how site-directed mutagenesis of crucial amino acid residues alter the structural and functional properties of this enzyme.

Crystal Structures at Different Resolutions Are Available

PA-LOX was first crystallized and X-rayed in 2013 and its 3D-structure was solved at a resolution of 1.75 Å (Garreta et al. 2013). In 2016 our group refined the structure to a resolution of 1.48 Å (Banthiya et al. 2016). The two structures are very similar and indicate that PA-LOX is a single domain protein, which lacks the N-terminal PLAT domain present in most eukaryotic LOXs (Figure 6.5). Instead PA-LOX involves a pair of long antiparallel helices of approximately 100 amino acid residues (Figure 6.5). These helices (shown in orange in Figure 6.5) have been postulated to function as lid for the entrance into the substrate-binding pocket (Garreta et al. 2013; Banthiya et al. 2016). Truncation variants of the enzyme lacking the N-terminal lid-structure were insoluble when expressed in *E. coli* and thus, these enzyme variants could not be functionally characterized. *N*-terminal proteolytic truncation of the lid-structure, which leads to the formation of a mini-PA-LOX variant, impaired the catalytic activity but improved substrate affinity (Lu et al. 2016) and these results are consistent with the lid-hypothesis.

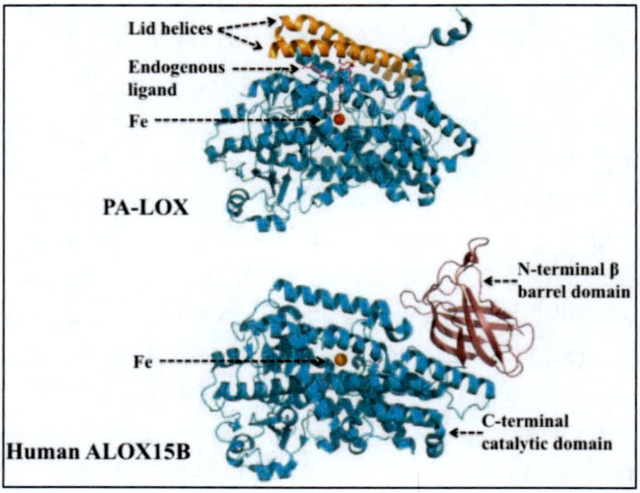

Figure 6.5. Structure of PA-LOX (PDB id: 5IR4) and human ALOX15B (PDB id: 4NRE).

The substrate-binding pocket of this enzyme is a bifurcated cavity, which involves a bacterial phospholipid molecule when the enzyme is expressed in *E. coli*. The sn1 fatty acid of this endogenous ligand occupies the non-heme iron-containing sub-cavity. The sn2 fatty acid is located in a second sub-cavity that does not have direct access to the iron. The polar head group of the endogenous phospholipid ligand occupies the joining lobby, which interconnects the two fatty acid binding sub-cavities (Garreta et al. 2013; Banthiya et al. 2016). The overall structure of the active site of PA-LOX indicates that this enzyme is capable of binding phospholipids and thus, it seems to be custom-made for oxygenation of complex ester or ether lipids carrying polyunsaturated fatty acids. However, the phospholipid oxygenase activity of PA-LOX is considerably lower than that of rabbit ALOX15, which involves a single cavity active site.

Reaction Specificity of PA-LOX

LOX are lipid-peroxidizing enzymes converting polyenoic fatty acids to their corresponding hydroperoxy derivatives. When arachidonic acid (a fatty acid with 20 carbon atoms and 4 isolated double bonds) is oxygenated non-enzymatically six different conjugated-diene containing hydroperoxides (5-HpETE, 8-HpETE, 9-HpETE, 11-HpETE, 12-HpETE, 15-HpETE) are formed. In these peroxides the number indicates the position of the carbon atom at which dioxygen is added during the oxygenation reaction. When LOXs are used as catalysts for the oxygenase reaction, only one out of the six different reaction products are formed. Arachidonic acid 5-LOX introduces molecular dioxygen at carbon atom 5 of the hydrocarbon backbone (formation of 5-HpETE). The other HpETE-isomers are only formed in small quantities. The principle capability of LOX to specifically synthesize one out of different possible product isomers is called reaction specifity.

Triad Concept

The reaction specificity of LOX-isoforms is one of the most interesting enzymatic properties of these enzymes and several concepts have been developed to identify amino acids controlling the reaction specificity of various LOX-isoforms. The positional specificity of mammalian ALOX15 orthologs can been explained by the Triad Concept (Borngraber and Browner, 1999; Vogel et al. 2010). According to this concept, a triad of amino acids, which align with Phe353 (Borngräber et al. 1996), Ile418/419 (Sloane et al. 1991; Sloane et al. 1995) and Ile 593 (Borngräber et al. 1999) of rabbit ALOX15 function as sequence determinants of ALOX15 specificity. Although distant in the primary structure, these residues are clustered in 3D structure of ALOX15 orthologs and form the bottom of the substrate-binding pocket. The bulkiness of their side chains determines the depth of this pocket. In 15-lipoxygenating ALOX15 orthologs, hydrogen abstraction takes place at C13 of the arachidonic acid backbone followed by oxygenation of C15 to form 15-HpETE. In contrast, in 12-lipoxygenating enzymes, hydrogen is abstracted from C10 of the arachidonic acid backbone followed by oxygenation of C12 to form 12-HpETE. When the triad determinants in a 15-lipoxygenating enzymes are replaced by less space-filling residues, the substrate fatty acid is able to penetrate deeper into the pocket so that C10 (instead of C13) is located in close proximity to the non-heme iron and this arrangement results in arachidonic acid 12-lipoxygenation. Similarly, introduction of more space-filling residues into a 12-lipoxygenating ALOX15 ortholog at these positions (mouse, rats, pigs) converts these proteins to 15-lipoxygenating enzymes (Vogel et al. 2010). All mammalian ALOX15 orthologs characterized so far (Adel et al. 2016; Suzuki et al. 1994; Vogel et al. 2010; Pekárová et al. 2015; Sloane et al. 1995; Borngräber et al. 1999) adhere to this concept regardless whether they are arachidonic acid 12- or 15-lipoxygenating. This concept is also applicable for mammalian ALOX12 orthologs but does not work for ALOX15B, ALOX12B orthologs. Applicability of this concept for PA-LOX has been tested but these data suggest that the Triad Concept may not be applicable for PA-LOX (Kalms et al. 2017). In contrast, the bacterial 12-lipoxygenating LOX-isoform from *M. xanthus* follows the Triad Concept. By dual sequence alignment with human ALOX15, the triad determinants of *M. xanthus* LOX were identified. Site-directed mutagenesis of Thr397 (Thr397Tyr) altered the reactions specificity of the enzyme as it oxygenated arachidonic acid mainly to 15-HpETE. In contrast, the major arachidonic acid oxygenation product of the wildtype enzyme was 12-HpETE. A similar alteration in reaction specificity was also observed for the oxygenation products of EPA and DHA. Interestingly, the reaction specificity of the Thr397Tyr mutant was not altered when C18 PUFAs such as alpha-linolenic acid or gamma-linolenic acid were used as substrates (An et al. 2018).

Jisaka Determinants

Human ALOX15B converts arachidonic acid to 15*S*-HpETE but the major arachidonic acid oxygenation of mouse Alox15b is 8*S*-HpETE. Similar to PA-LOX, mammalian ALOX15B orthologs do not follow the Triad Concept and thus, the triad determinants do not impact the reaction specificity of these enzymes. In 2000, Jisaka and his colleagues reported for human ALOX15B that Glu604 and Lys605 appear to function as sequence determinants for the reaction specificity of this enzyme. Mutation of these amino acids (Jisaka determinants) to the corresponding residues present at these positions in murine Alox15b converted the 15-lipoxygenating human ALOX15B to an 8-lipoxygenating enzyme (Jisaka et al. 2000). To test this hypothesis for the arachidonic acid 15-lipoxygenating PA-LOX, we mutated the corresponding amino acids of PA-LOX to residues present at these positions in mouse Alox15b. We generated the Glu604Tyr+Lys605His double mutant but this enzyme variant did not show any alteration in the reaction specificity (Banthiya et al. 2016). Thus, the Jisaka determinants do not play a major role for the reaction specificity of PA-LOX.

Ala-vs-Gly Concept (Coffa/Brash Concept)

The enantioselectivity of many LOX-isoforms is controlled by a single amino acid localized at a critical position in the primary structure. Multiple sequence alignment of a number of eukaryotic LOXs suggested that most *S*-lipoxygenating enzymes contain an alanine at this critical position (Coffa site). In contrast, *R*-lipoxygenating enzymes contain a glycine at the corresponding position. When this Ala in human ALOX15B was mutated to Gly, the product profile was shifted from 15*S*-HpETE formation to dominant 11*R*-HpETE formation. Similarly, when the Gly of human ALOX12B was mutated to Ala, the product profile was shifted from dominant 12*R*-HpETE formation to major 8*S*-lipoxygenation. Although this concept works well with several LOX-isoforms it is only partly applicable to most mammalian ALOX15 orthologs and for human ALOX12 (Vogel et al. 2010). Moreover, it does not explain the reaction specificity of zebrafish LOX1 (Jansen et al. 2011). The 11*S*-LOX from *M. xanthus* has an alanine (Ala409) at its Coffa site and it catalyses the oxygenation of AA to 11*S*-HETE (An and Oh, JLR 2018, in press). When this alanine was mutated to a glycine, the mutant enzyme forms small amounts 15*R*-HpETE in addition to 11*S*-HpETE. However, the major arachidonic acid oxygenation product still remained 11*S*-HpETE (An and Oh, JLR 2018, in press).

PA-LOX is an *S*-lipoxygenating enzyme and multiple sequence alignments indicated that it carries an alanine (Ala420) at its Coffa site (Coffa & Brash 2004). Ala420Gly exchange induced alterations in positional specificity and the enantioselectivity of the enzyme as predicted by the Coffa-concept. However, as for mammalian ALOX15 orthologs (Vogel et al. 2010) and for the *M. xanthus* 11*S*-LOX (An & Oh 2018) the alterations were incomplete. The Ala420Gly mutant converted arachidonic acid to 11*R*-HETE and 15*S*-HETE in a ratio of 1.5:1. These data suggest that PA-LOX partly adheres to the Ala vs. Gly concept (K

alms et al. 2017). When the Ala of the 9*R*-LOX from the cyanobacterium *Nostoc sp. PCC7120* was mutated to a less bulky Gly, no significant alteration in the reaction specificity was observed. However, when bulkier residues, such as Val or Ile, were introduced at this position 13*S*-HpODE was the major oxygenation product (Andreou et al. 2008). Unfortunately, among the canonical amino acids there are no smaller amino acids than Gly. Thus, it could not be tested whether the introduction of an even smaller residue than Gly at position 420 into PA-LOX would lead to more complete changes in the reaction specificity of this enzyme.

In an attempt to explain the structural basis for the altered reaction specificity observed for Ala420Gly mutant of PA-LOX, the mutant enzyme was crystallized to a resolution of 1.8 Å (Kalms et al. 2017). Although the overall 3 dimensional structures of the wildtype (Banthiya, 2016) and the mutant enzyme (Kalms et al. 2017) were similar, the protein tunneling software Caver 3.0 (Chovancova et al. 2012) suggested an alternative path for intra-enzyme oxygen diffusion in the mutant enzyme. Structural modeling of the enzyme-arachidonic acid complex suggested that dioxygen molecules are directed to C15 of the arachidonic acid backbone in wildtype-PA-LOX. However, Ala420Gly exchange redirected intra-enzyme oxygen diffusion. In fact, Ala420Gly exchange bifurcated the putative oxygen diffusion channel so that molecular dioxygen can reach both C11 and C15 of the arachidonic acid backbone explaining the dual reaction specificity of the mutant enzyme (Kalms et al. 2017). These structural data suggest that removal of the CH3-group of Ala420 opened a side channel, which is blocked in the wildtype enzyme. Thus, for the mutant PA-LOX atmospheric dioxygen can also reach the 11*R*- position of arachidonic acid in addition to the 15*S*-position. These structural conclusions were consistent with the alterations of reaction kinetic properties of the enzyme. We found that the Ala420Gly mutant exhibited a 2-fold higher oxygen affinity [Km(O_2) = 235 μM] than wildtype-PA-LOX [Km(O_2) = 406 μM]. In other words, following the newly opened path of oxygen diffusion it was easier for oxygen molecules to reach the site of oxygen insertion into the fatty acid substrate (Kalms et al. 2017).

The Clamp-Model of Reaction Specificity

The CspLOX2 from the cyanobacterium *Cyanothece PCC8801* is an iron containing LOX, which converts linoleic acid to the bisallylic (11R)-hydroperoxy derivative. Its crystal structure was recently solved at a resolution of 1.8 Å and multiple mutagenesis studies suggested that several catalytically important residues lining the active site of the enzyme direct molecular dioxygen to a preferred position of the substrate fatty acid (Newie et al. 2017). From their experimental data the authors concluded that an active site clamp that arrests the pentadiene system of the substrate fatty acid. In addition, there appears to be a targeted intra-enzyme oxygen diffusion and these two factors control the stereochemistry of the LOX reaction. Unfortunately, corresponding mutations have not been carried out for PA-LOX and thus, it remains unclear whether this mechanistic concept may be applicable for this LOX-isoform.

Possible Biological Relevance of PA-LOX

Although a number of bacterial LOX-isoforms have been extensively characterized in the past little is known on the biological relevance of these enzymes for bacterial physiology. This lack of knowledge is probably related to the limited distribution of these enzymes in bacteria and also to the fact that bacteria in general do not need polyenoic fatty acid metabolizing enzymes for survival (Oliver & Colwell 1973). On the other hand, some species of different bacterial families, such as *Pseudomonaceae*, *Myxococcocea*e and *Cyanobacteria*, have acquired LOX genes from higher eukaryotes *via* horizontal gene transfer and have retained them throughout evolution (Porta & Rocha-Sosa 2001). Thus, there must by an evolutionary pressure on these genes. In other words, LOX-expressing bacteria should have an evolutionary advantage over LOX-deficient individuals of the same species. Unfortunately, for the time being it has not been explored what this evolutionary advantage might be. Although there is no unifying concept for the biological role of LOX-isoforms in bacteria, there are a number of hypotheses for the bioactivity of PA-LOX and these concepts are summarized in Figure 6.6. Each of these hypotheses will be discussed in more detail in the next sections of this review.

Hypothesis 1: Expression of PA-LOX Increases the Invasive Capacity of PA

PA is an invasively growing pathogen, which kills host cells when replicating in the host (Rejman et al. 2007). In 2013, Garreta and colleagues (Garreta et al. 2013) postulated that PA-LOX expression might increase the invasive capacity of *PA*. To test this hypothesis, they performed *in vitro* cytotoxicity and invasiveness assays using a PA-LOX expressing strain of *PA* and a PA-LOX-deficient strain, which was created by targeted gene manipulation. The authors observed a 50% reduction of the invasiveness for PA-LOX deficient bacteria (Garreta et al. 2013). From these data, the authors concluded that PA-LOX might play an essential role in altering host cell membrane integrity when interacting with eukaryotic cells. If this hypothesis is true PA-LOX might play a role for the pathogenicity of this bacterium. However, when testing the cytotoxicity of LOX-sufficient and LOX-deficient *PA* strains the authors did not observe an increased LDH released for the PA-LOX sufficient *PA* strain (Garreta et al. 2013).

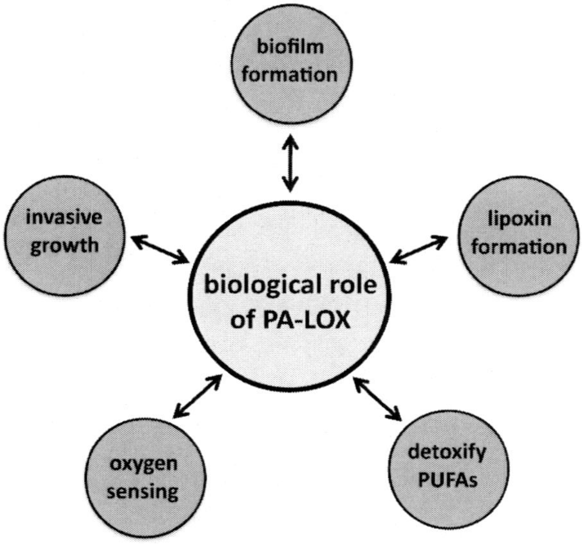

Figure 6.6. The possible biological roles of PA-LOX.

As indicated in previous sections, PA-LOX is capable of oxygenating phospholipids and biomembranes in different subcellular *in vitro* model system. Next, we explored the ability of the enzyme to oxygenate the membrane lipids of intact cells. Since *PA* frequently infects the lungs (De Bentzmann & Plésiat 2011) and sometimes also occurs in the blood (*PA* sepsis) we employed human erythrocytes (Banthiya et al. 2015) and cultured human lung epithelial cells (A549 cells) as cellular targets (Aldrovandi et al. 2018). When recombinant PA-LOX was incubated *in vitro* with intact human erythrocytes (RBCs) strong hemolysis was induced during a 24 h incubation period. In fact, at the end of the incubation period about 50% of the cells were disrupted as indicated by the release of hemoglobin into the cell free supernatant. In control incubations with purified rabbit ALOX15 (same arachidonic acid oxygenase activity) only 1-2% of the cells were lyzed (Banthiya et al. 2015). The oxidized phospholipids formed during these *in vitro* incubations were analyzed using a lipidomic approach. LC-MS/MS analysis of the lipid extracts prepared from the incubation mixtures indicated a complex pattern of oxidation products involving 40 different phosphatidylethanolamine and phosphatidylcholine species carrying oxidized fatty acid, such as hydroxylated octadecadienoic acids, hydroxy- and keto-eicosatetraenoic acids and hydroxylated docosahexaenoic acid. In addition to these derivatives, we also detected oxygenation products of less frequently occurring polyenoic fatty acids such as adrenic acid, docosatetraenoic acid and docosapentaenoic acid (Aldrovandi et al. 2018).

A similar profile of oxygenation product was observed when cultured lung epithelial cells (A549cells) were employed as cellular targets. However, with A549 cells the OH-PUFA/PUFA ratio, which may be considered a suitable measure for the degree of oxygenation of the membrane lipids, was ~3.7% when 250 µg PA-LOX/ml was used. In side-by-side incubations this ratio was ~19% when 70 µg of pure PA-LOX/ml was incubated with RBC. These data suggested that the membrane lipids of A549 cells are less effectively oxygenated by PA-LOX when compared with the membrane lipids of RBC. It is important to note that much more RBCs were present in the incubation samples than A549 cells (10^9 RBCs *vs.* 5×10^6 A549 cells) and thus, each RBC was loaded with less amounts of PA-LOX when compared with A549 cells (Aldrovandi et al. 2018).

The most striking difference between the incubations of the two cell types was that RBCs were strongly disrupted whereas A549 cells remained morphologically intact (Aldrovandi et al. 2018). Although we observed an impaired capacity for trypan blue exclusion, there was no sign of cellular lysis of A549 cells. In fact, live imaging of pre-confluent A549 cells even indicated a higher mitosis rate of PA-LOX treated A549 cells (Aldrovandi et al. 2018). The molecular reasons for the cell specific effects have not been explored in detail but they may be related to the different fatty acid composition of the membrane lipids of these two cell types but also to more efficient protective repair mechanisms in A549 cells. Although erythrocytes involve effective anti-oxidative defense systems they have a reduced capacity to remove oxygenated fatty acids from the membrane lipids and to metabolize oxidized fatty acids. RBC do neither contain mitochondria nor peroxisomes and thus ß-oxidation of oxidized fatty acids is not possible Although hydrolysis-esterification cycles have been reported for RBCs such metabolic events are energy consuming and RBCs have only a compromised energy metabolism. Thus, their repair capacity of the membrane phospholipid bilayer is limited so that external oxidative challenge may lead to premature hemolysis. Taken together, these data indicate that PA-LOX is capable of oxidizing the membrane phospholipids of eukaryotic cells indicating that this enzyme might function as a virulence factor. Thus, patients suffering from *PA* infections may benefit from pharmacological intervention with PA-LOX inhibitors.

Hypothesis 2: PA-LOX Promotes Biofilm Formation

Microbes form biofilms in response to various factors, which include cellular recognition of attachment sites, nutritional cues or exposure of planktonic cells to sub-inhibitory concentrations of antibiotics. A cell that switches from the planktonic mode of existence to biofilm mode of growth undergoes a phenotypic shift, in which the expression of large clusters of genes are either up- or down-regulated regulated (An & Parsek 2007). Biofilms are complex biological systems, in which bacteria organize themselves into functional communities, which may include a single species or even diverse groups of microorganisms. In biofilms bacteria can share nutrients and are sheltered from harmful factors of the environment, such as desiccation, antibiotics, and the host body's immune system. Thus the formation of biofilms may be regarded as evasion strategy, which helps bacteria to survive, when the immune cells of the infected host attack them.

PA-LOX has been implicated in the formation of biofilms by PA (Deschamps et al. 2016). To test this hypothesis the PA strain PAO1 that expresses PA-LOX and a transposon knockout strain of PAO1 were grown on the surface of human lung airway epithelial cells. In this biotic biofilm assays the authors observed a less effective biofilm formation with the PA strain that does not express PA-LOX (Deschamps et al. 2016). The impaired biofilm formation was compensated when recombinant PA-LOX was added to the assay system and when expression of PA-LOX was induced in the PA-LOX deficient strain. In order to further strengthen their hypothesis the authors excluded, that reduction in biofilm formation of PA-LOX deficient bacteria had anything to do with an altered attachment of the bacteria to the airway epithelial cells. In negative control experiments it was observed that PA-LOX expression was not required for biofilm formation on abiotic surfaces (Deschamps et al. 2016). Taken together, these data suggest that PA-LOX might play a role in biofilm formation on biotic surfaces. Although very interesting, this conclusion needs further experimental support. Biofilms are formed on biotic and abiotic surfaces by a large number of bacterial species and most of them do not carry LOX genes. For instance, *E.coli* are frequently employed as model systems to study the mechanistic basis of biofilm formation (Beloin et al. 2008) but the *E. coli* reference genome does not contain functional LOX genes. However, it may well be that different bacterial species developed alternative strategies for biofilm formation and that PA-LOX is involved in those mechanisms evolved in PA.

Hypothesis 3: PA-LOX Down-Regulates The Immune Response of The Host By The Formation of Pro-Resolving Lipoxins

Mammalian ALOX15 orthologs (man, rabbit, pig, mouse) are capable of oxygenating arachidonic acid to anti-inflammatory and pro-resolving mediators called lipoxins (Romano 2010). When we studied the capability of PA-LOX to synthesize such lipid mediators (Banthiya et al. 2016), we found that recombinant PA-LOX converted 5*S*-HETE and 5*S*,6*S*/*R*-DiHETE to lipoxin A and lipoxin B isomers. These compounds down-regulate the inflammatory response inducing inflammatory resolution (Romano 2010). If one translates our *in vitro* findings into the *in vivo* situation one might speculate that the biosynthesis of anti-inflammatory lipoxins by PA-LOX will silence the hosts immune reaction intended to fight the pathogen. If this hypothesis is true PA-LOX expression is part of an evasion strategy of *PA* to circumvent and even down regulate the immune system of the host. This evasion strategy should clearly confer an evolutionary advantage to PA-LOX expressing *PA* strains.

When recombinant PA-LOX was characterized in a parallel study, no lipoxin synthase activity of the enzyme was detected (Deschamps et al. 2016) and these findings contradict the results reported in our study (Banthiya et al. 2016). In the Deschamps study the authors employed 15-HpETE, 15-HETE, 5-HpETE, and 5-HETE as substrates for lipoxin formation. The major reason why the lipoxin synthase activity escaped detection in the Deschamps was that the authors employed in their lipoxin synthase activity assays enzyme concentrations, which were sufficiently high to follow arachidonic acid oxygenation but were not high enough to induce lipoxin formation. From previous studies on the lipoxin synthase activity of mammalian LOX we knew that for reliable assays of lipoxin formation at least 10-fold higher enzyme concentrations are needed (Kuhn et al. 1987). In fact, when we employed recombinant PA-LOX at concentrations usually used for arachidonic acid oxygenation we did not detect any lipoxin synthase activity. Thus, there is no real contradiction between the two reports; the experimental conditions for the assay systems were simply different. It should, however, be stressed that from an *in vivo* perspective, two points needs to be clarified: i) Are such high amounts of PA-LOX actually reached *in vivo* during a *PA* infection? ii) Are the amounts of lipoxins synthesized by PA-LOX during the time course of a *PA* infection high enough to silence the hosts' immune response? To answer these questions *in vivo* studies on *PA* infected patients should be carried out.

Hypothesis 4: PA-LOX Is Part of The Oxygen Sensing System

Atmospheric dioxygen functions as second substrate for all LOX-isoforms. The oxygen affinity of most mammalian LOXs varies between 3-30 μM (Yamamoto et al. 2002). However, PA-LOX exhibits a very high oxygen Km of more than 400 μM (Kalms et al. 2017) and this data indicates that this enzyme has an unusually low affinity for oxygen. This kinetic peculiarities means that under physiological conditions most mammalian LOX isoforms will work in the range of oxygen saturation and that minor alterations in the local oxygen concentrations will hardly impact the reaction rate. In contrast, under physiological conditions, PA-LOX is not oxygen saturated and thus, even minor alterations in the local oxygen concentration will strongly impact the reaction rate. Since such kinetic properties are characteristic for sensor proteins (Berra 2003) PA-LOX might be involved in oxygen sensing. At low oxygen concentrations the LOX reaction proceeds with a low rate, which means that the cellular peroxide content is low. In contrast, at higher oxygen concentrations the LOX reaction proceeds with higher rates, which increases the cellular peroxide levels. Thus, PA-LOX may translate the existing oxygen concentrations in a proportional way into alterations in the cellular redox state and these alterations will impact the expression pattern of redox sensitive genes. In this way the cellular metabolism can be adapted to the existing local oxygen concentrations. Although there is currently no experimental proof for this scenario, the unusual kinetic properties of PA-LOX may implicate this enzyme in oxygen sensing, which adapts the cellular metabolism to the existing oxygen concentrations.

Hypothesis 5: PA-LOX Contributes To Detoxification of Polyenoic Fatty acids

Unsaturated fatty acids, which hardly occur in prokaryotes, exhibit toxic effects for many bacteria (Raychowdhury et al. 1985; Greenway & Dyke 1979). During evolution bacteria developed defense mechanisms to metabolize such toxins. For instance, oleic acid is toxic for *Streptococcus pyogenes M49*, a human pathogenic bacterium. However, this bacterium expresses a fatty acid double bond hydratase that metabolizes oleic acid to a hydroxylated derivative (Volkov et al. 2010), which is not toxic any more. Although recombinant PA-LOX does not oxygenate oleic acid it might contribute to detoxification of PUFAs, which are suitable substrates for this enzyme.

FUTURE PROSPECTS

Although PA-LOX has been discovered more than 15 years ago the biological relevance of the enzyme has not been well characterized. To improve our understanding on the *in vivo* functionality of PA-LOX research tools must be constructed, which can be employed to explore to biological role of this enzyme. Genetically modified PA-strains lacking the PA-LOX genes are available but their functional characterization is still pending. In the near future we expect additional research along three different lines, which will help to shed light on the biological relevance of this enzyme

Tool Development

Not all *PA* strains, which have been isolated in the past, carry the PA-LOX gene. There are PA-LOX deficient strains and it would be of clinical interest to determine whether a novel clinical *PA* isolate is PA-LOX sufficient or PA-LOX deficient. For this purpose a simple PCR-based genotyping assay should be developed, which employs PA-LOX specific primers. For primer design, regions in the cDNA should be selected which show a high degree of sequence conservation. For LOXs the most strongly conserved sequence regions are found around the two iron ligand clusters (N-terminal iron binding cluster, His-Leu-Leu-Arg-Gly-His; C-terminal iron cluster, His-Ala-Ala-Val-Asn) and thus, such primers should flank at least one of the two iron ligand cluster to make sure that these functionally important parts of the PA-LOX gene are present in the genome of the clinical isolates. If this is not the case or if sequencing of the PCR product reveals corrupted sequences, the isolate is functionally PA-LOX deficient.

Recombinant expression and purification of PA-LOX yields large amounts of pure enzyme and this protein can be employed to prepare poly- and/or monoclonal antibodies. Such antibodies should be helpful to explore the cellular and subcellular localization of PA-LOX in *PA* infected patients. Moreover, such antibodies can be used to detect PA-LOX in clinical isolates of *PA* to discriminate between PA-LOX expressing and PA-LOX deficient strains. Finally, such antibodies might even be helpful to quantify the amounts of *PA* present in the blood in patients suffering from *PA*-sepsis.

In Vivo Data Characterizing The Biological Role of PA-LOX

Although PA-LOX has been extensively characterized with respect to its protein-chemical and enzymatic properties there are hardly any *in vivo* data characterizing the biological relevance of this enzyme. From the *in vitro* interaction of recombinant PA-LOX with human erythrocytes we know, which products are formed during pathogen-RBCs interaction. It would be interesting to explore whether such product can also be detected in red blood cell membranes when erythrocytes are incubated *in vitro* with *PA* bacteria expressing PA-LOX. As suitable controls PA-LOX deficient strains or genetically modified *PA* strains should be employed. According to our current knowledge PA-LOX is secreted from the bacteria and should be able to directly attack RBC membranes. Thus, a similar pattern of oxygenation products in the erythrocyte membranes is expected when PA-LOX expressing bacteria are incubated with isolated RBCs. Before going into *in vivo* studies such more complex *in vitro* incubations should be carried out.

In addition to these more complex *in vitro* studies should be tested whether similar oxygenation products can be detected in patients suffering from chronic and/or acute *PA* infection. If one translates our *in vitro* data into the *in vivo* situation one would expect that RBCs of *PA*-patients involve specific oxygenation products. The most ideal candidates for such experiments are patients with clinically manifested *PA*-sepsis, in which the *PA* bacteria have been detected in the blood. Unfortunately, such patients are rather rare so that patients with local *PA*-infections (lungs) might be analyzed. However, it might well be that PA-LOX damaged erythrocytes are selectively removed from the circulation so that the average degree of the oxygenated fatty acids in the red cell membranes is similar or even lower than normal.

PA-LOX Inhibitors as Putative Drugs

The ability of *PA* to develop an intrinsic resistance against antibiotics makes it difficult to treat fatal infections caused by this bacterium Therefore, many studies are currently focused on identifying novel virulence factors for this bacterium (Hurley et al. 2012). Pharmacological interference with novel virulence factors might complement or even substitute conventional antibiotic therapies. Since PA-LOX is a candidate for such a novel virulence factor the development of specific PA-LOX inhibitors might be of clinical relevance. A study working with recombinant PA-LOX screened the library of Pharmacologically Active Compounds (LOPAC) using a high-throughput screening assay to search for PA-LOX specific inhibitors. The IC_{50} values of the most effective compounds were in the one-digit μM range. The most potent compound exhibited an IC_{50} of 4.1 ± 1.4 μM. Unfortunately, this compound did also inhibit human ALOX12 and human ALOX15B (Deschamps et al. 2016) and thus its isoform specificity is somewhat limited. We tested nordihydroguaiaretic acid (NDGA), which is standard LOX inhibitor with rather low isoform specificity (it inhibits virtually all LOX isoforms tested so far), as inhibitor of PA-LOX and found an IC_{50} value of 30 μM (Banthiya et al., unpublished data). Because of the relatively high IC_{50}, the low isoform specificity and the anti-oxidative properties NDAG might not be well suited as starting point of an optimization strategy.

If one compares the 3D-structures of various pro- and eukaryotic LOX-isoforms PA-LOX is somewhat unique because of two structural peculiarities: i) It is a one-domian enzyme, which lacks to N-terminal ß-barrel domain of most mammalian LOX isoforms. This structural peculiarity might not be helpful for inhibitor development. ii) The substrate-binding pockets of all LOX-isoforms X-rayed so far are internal cavities, which are lined by hydrophobic amino acids and involve the non-heme iron. For some LOX-isoforms these cavities are connected with the protein surface so that the fatty acid substrate can easily diffuse into the active site. For other isoforms these connections are corked by bulky amino acids (FY-cork for human ALOX5) but the de-corking mechanisms, which are required for effective substrate binding, are not well understood. PA-LOX involves a bifurcated substrate-binding pocket and a joining lobby that interconnects the two subcavities. The lobby has access to the protein surface but the entrance is blocked in part by the helical lid structure. Upon movement of the lid helices a phospholipid can penetrate into the substrate-binding pocket and its two fatty acid moieties are separately bound in the two sub-cavities. The polar head group of the phospholipid molecule will occupy the lobby region. Such a structural arrangements have not been shown for any other LOX-isoform. Thus, this structural peculiarity might be helpful to identify low molecular weight compounds, which optimally fit as active-site probes into the bifurcated substrate-binding pocket of PA-LOX but not into the active site of other LOX-isoforms.

CONCLUSION

Recombinant PA-LOX is capable of oxidizing the membrane phospholipids of host cells and this functional property may improve the invasive characteristics of *PA*. This review summarized the enzymatic properties of PA-LOX and discussed the biological roles this enzyme may play when *PA* infects a host organism. Unfortunately, little information is currently available about the *in vivo* relevance of the reported *in vitro* effects. To change this situation more detailed *in vivo* studies are needed to explore of whether or not PA-LOX might constitute a molecular target for future anti-*PA* therapy. Since *PA* infections challenge the health care systems of both developed and developing countries there is an urgent need for international collaborative research strategies, which should include basic researchers of different disciplines (microbiologists, biochemists, structural biologists, chemists) as well as clinicians, to explore whether patients suffering from PA-infections might benefit from anti-PA-LOX therapy.

REFERENCES

Aanaes, K. et al. (2015). Sinus biofilms in patients with cystic fibrosis: is adjusted eradication therapy needed? *European archives of oto-rhino-laryngology : official journal of the European Federation of Oto-Rhino-Laryngological Societies (EUFOS) : affiliated with the German Society for Oto-Rhino-Laryngology - Head and Neck Surgery*, 272(9), pp. 2291–7.

Adel, S. et al. (2016). Evolutionary alteration of ALOX15 specificity optimizes the biosynthesis of antiinflamm1. Adel, S. et al. Evolutionary alteration of ALOX15 specificity optimizes the biosynthesis of antiinflammatory and proresolving lipoxins. *Proceedings of the National Academy of Sciences of the United States of America*, 113(30), pp. E4266-4275.

Aldrovandi, M. et al. (2018). Specific oxygenation of plasma membrane phospholipids by *Pseudomonas aeruginosa* lipoxygenase induces structural and functional alterations in mammalian cells. *Biochimica et Biophysica Acta*, 1863(2), pp. 152–164.

An, D. & Parsek, M.R. (2007). The promise and peril of transcriptional profiling in biofilm communities. *Current Opinion in Microbiology*, 10(3), pp. 292–6.

An, J.-U., Hong, S.-H. & Oh, D.-K. (2018). Regiospecificity of a novel bacterial lipoxygenase from *Myxococcus xanthus* for polyunsaturated fatty acids. *Biochimica et Biophysica Acta*, 1863(8), pp. 823–833.

An, J.-U. & Oh, D.-K. (2018). Stabilization and improved activity of arachidonate 11S-lipoxygenase from proteobacterium *Myxococcus xanthus*. *Journal of Lipid Research*.

Andreou, A. et al. (2008). Properties of a mini 9R-lipoxygenase from Nostoc sp. PCC 7120 and its mutant forms. *Phytochemistry*, 69(9), pp. 1832–1837.

Banthiya, S. et al. (2015). Secreted lipoxygenase from *Pseudomonas aeruginosa* exhibits biomembrane oxygenase activity and induces hemolysis in human red blood cells. *Archives of Biochemistry and Biophysics*, 584, pp. 116–124.

Banthiya, S. et al. (2016). Structural and functional basis of phospholipid oxygenase activity of bacterial lipoxygenase from *Pseudomonas aeruginosa*. *Biochimica et Biophysica Acta (BBA) - Molecular and Cell Biology of Lipids*, 1861(11), pp. 1681–1692.

Belkner, J. et al. (1991). The oxygenation of cholesterol esters by the reticulocyte lipoxygenase. *FEBS Letters*, 279(1), pp. 110–114.

Beloin, C., Roux, A. & Ghigo, J.M. (2008). *Escherichia coli* biofilms. *Current Topics in Microbiology and Immunology*, 322, pp. 249–289.

Ben Haj Khalifa, A. et al. (2011). Virulence factors in *Pseudomonas aeruginosa*: mechanisms and modes of regulation. *Annales de Biologie Clinique*, 69(4), pp. 393–403.

De Bentzmann, S. & Plésiat, P. (2011). The *Pseudomonas aeruginosa* opportunistic pathogen and human infections. *Environmental Microbiology*, 13(7), pp. 1655–1665.

Berra, E. (2003). HIF prolyl-hydroxylase 2 is the key oxygen sensor setting low steady-state levels of HIF-1 in normoxia. *The EMBO Journal*, 22(16), pp. 4082–4090.

Borngräber, S. et al. (1996). Phenylalanine 353 is a Primary Determinant for the Positional Specificity of Mammalian 15-Lipoxygenases. *Journal of Molecular Biology*, 264(5), pp. 1145–1153.

Borngräber, S. et al. (1999). Shape and specificity in mammalian 15-lipoxygenase active site. The functional interplay of sequence determinants for the reaction specificity. *Journal of Biological Chemistry*, 274(52), pp. 37345–37350.

Brandel, J. et al. (2012). Pyochelin, a siderophore of *Pseudomonas aeruginosa*: physicochemical characterization of the iron(III), copper(II) and zinc(II) complexes. *Dalton transactions (Cambridge, England : 2003)*, 41(9), pp. 2820–34.

Brash, A.R. (1999). Lipoxygenases: occurrence, functions, catalysis, and acquisition of substrate. *The Journal of Biological Chemistry*, 274(34), pp. 23679–23682.

Brodhun, F. et al. (2009). Identification of PpoA from *Aspergillus nidulans* as a fusion protein of a fatty acid heme dioxygenase/peroxidase and a cytochrome P450. *The Journal of Biological Chemistry*, 284(18), pp. 11792–805.

Busquets, M. et al. (2004). Isolation and characterization of a lipoxygenase from *Pseudomonas* 42A2 responsible for the biotransformation of oleic acid into (S)-(E)-10-hydroxy-8-octadecenoic acid. *Antonie van Leeuwenhoek, International Journal of General and Molecular Microbiology*, 85(2), pp.129–139.

Chovancova, E. et al., 2012. CAVER 3.0: a tool for the analysis of transport pathways in dynamic protein structures. *PLoS Computational Biology*, 8(10), p.e1002708.

Clifton, I.J. & Peckham, D.G., 2010. Defining routes of airborne transmission of *Pseudomonas aeruginosa* in people with cystic fibrosis. *Expert Review of Respiratory Medicine*, 4(4), pp.519–29.

Coffa, G. & Brash, A.R., 2004. A single active site residue directs oxygenation stereospecificity in lipoxygenases: stereocontrol is linked to the position of oxygenation. *Proceedings of the National Academy of Sciences of the United States of America*, 101(44), pp.15579–15584.

Deschamps, J.D. et al., 2016. Biochemical and Cellular Characterization and Inhibitor Discovery of *Pseudomonas aeruginosa* 15-Lipoxygenase. *Biochemistry*, 55(23), pp.3329–3340.

Funk, C.D. et al., 2002. Lipoxygenase genes and their targeted disruption. *Prostaglandins & Other Lipid Mediators*, 68–69, pp.303–312.

Funk, M.O. et al., 1990. Role of iron in lipoxygenase catalysis. *Journal of the American Chemical Society*, 112(13), pp.5375–5376.

Garreta, A. et al., 2013. Structure and interaction with phospholipids of a prokaryotic lipoxygenase from *Pseudomonas aeruginosa*. *FASEB Journal*, 27(12), pp.4811–4821.

Garreta, A., Carpena, X. & Busquets, M., 2011. Crystallization of the lipoxygenase of *Pseudomonas aeruginosa* 42A2, evolution and phylogenetic study of the subfamilies of the lipoxygenases. *Recent Advances in Pharmaceutical Sciences (Cap. 11)*, 661(2), pp.247–273.

Gellatly, S.L. & Hancock, R.E.W., 2013. *Pseudomonas aeruginosa*: New insights into pathogenesis and host defenses. *Pathogens and Disease*, 67(3), pp.159–173.

Greenway, D.L. & Dyke, K.G., 1979. Mechanism of the inhibitory action of linoleic acid on the growth of *Staphylococcus aureus*. *Journal of General Microbiology*, 115(1), pp.233–45.

Groisman, E.A., 2001. The pleiotropic two-component regulatory system PhoP-PhoQ. *Journal of Bacteriology*, 183(6), pp.1835–42.

Guerrero, A. et al., 1997. Oxydation of oleic acid to (E)-10-hydroperoxy-8-octadecenoic and (E)-10-hydroxy-8-octadecenoic acids by *Pseudomonas* sp. 42A2. *Biochimica et Biophysica Acta*, 1347(1), pp.75–81.

Haeggström, J.Z. & Funk, C.D., 2011. Lipoxygenase and leukotriene pathways: biochemistry, biology, and roles in disease. *Chemical Reviews*, 111(10), pp.5866–5898.

Hansen, J. et al., 2013. Bacterial lipoxygenases, a new subfamily of enzymes? A phylogenetic approach. *Applied Microbiology and Biotechnology*, 97(11), pp.4737–4747.

Hauser, A.R. & Ozer, E.A., 2008. Pseudomonas aeruginosa: an uninvited guest refuses to leave. *American Journal of Respiratory And Critical Care Medicine*, 178(5), pp.438–439.

Horn, T. et al., 2015. Evolutionary aspects of lipoxygenases and genetic diversity of human leukotriene signaling. *Progress in Lipid Research*, 57, pp.13–39.

Hurley, M.N., Cámara, M. & Smyth, A.R., 2012. Novel approaches to the treatment of *Pseudomonas aeruginosa* infections in cystic fibrosis. *European Respiratory Journal*, 40(4), pp.1014–1023.

Iny, D. et al., 1993. Lipoxygenase of Thermoactinomyces vulgaris, purification and characterization of reaction products. *International Journal of Biochemistry*, 25(9), pp.1313–1323.

Jansen, C. et al., 2011. Stereocontrol of arachidonic acid oxygenation by vertebrate lipoxygenases: newly cloned zebrafish lipoxygenase 1 does not follow the Ala-versus-Gly concept. *The Journal of Biological Chemistry*, 286(43), pp.37804–37812.

Jisaka, M. et al., 2000. Identification of Amino Acid Determinants of the Positional Specificity of Mouse 8S-Lipoxygenase and Human 15S-Lipoxygenase-2. *Journal of Biological Chemistry*, 275(2), pp.1287–1293.

Kalms, J. et al., 2017. The crystal structure of Pseudomonas aeruginosa lipoxygenase Ala420Gly mutant explains the improved oxygen affinity and the altered reaction specificity. *Biochimica et Biophysica Acta (BBA) - Molecular and Cell Biology of Lipids*, 1862(5), pp.463–473.

Kang, C.-I. et al., 2003. *Pseudomonas aeruginosa* Bacteremia: Risk Factors for Mortality and Influence of Delayed Receipt of Effective Antimicrobial Therapy on Clinical Outcome. *Clinical Infectious Diseases*, 37(6), pp.745–751.

Kaplan, J.B., 2010. Biofilm dispersal: mechanisms, clinical implications, and potential therapeutic uses. *Journal of Dental Research*, 89(3), pp.205–18.

Kariminik, A., Baseri-Salehi, M. & Kheirkhah, B., 2017. *Pseudomonas aeruginosa* quorum sensing modulates immune responses: An updated review article. *Immunology Letters*, 190, pp.1–6.

Kirkland, T.N. et al., 1998. Evaluation of the proline-rich antigen of Coccidioides immitis as a vaccine candidate in mice. *Infection and Immunity*, 66(8), pp.3519–3522.

Koeduka, T., Kajiwara, T. & Matsui, K., 2007. Cloning of lipoxygenase genes from a cyanobacterium, Nostoc punctiforme, and its expression in *Eschelichia coli*. *Current Microbiology*, 54(4), pp.315–319.

Kominos, S.D., Copeland, C.E. & Grosiak, B., 1972. Mode of transmission of *Pseudomonas aeruginosa* in a burn unit and an intensive care unit in a general hospital. *Applied Microbiology*, 23(2), pp.309–12.

Kuhn, H. et al., 1987. Formation of lipoxin B by the pure reticulocyte lipoxygenase via sequential oxygenation of the substrate. *European Journal of Biochemistry*, 169(3), pp.593–601.

Kutzner, L. et al., 2017. Mammalian ALOX15 orthologs exhibit pronounced dual positional specificity with docosahexaenoic acid. *Biochimica et Biophysica Acta*, 1862(7), pp.666–675.

Lang, I. et al., 2008. A lipoxygenase with linoleate diol synthase activity from *Nostoc* sp. PCC 7120. *Biochemistry Journal*, 410, pp.347–357.

Lang, I. & Feussner, I., 2007. Oxylipin formation in *Nostoc punctiforme* (PCC73102). *Phytochemistry*, 68(8), pp.1120–7.

Lu, X. et al., 2013. Overproduction, purification, and characterization of extracellular lipoxygenase of *Pseudomonas aeruginosa* in *Escherichia coli*. *Applied Microbiology and Biotechnology*, 97(13), pp.5793–5800.

Lu, X. et al., 2016. The N-Terminal α-Helix Domain of *Pseudomonas aeruginosa* Lipoxygenase Is Required for Its Soluble Expression in *Escherichia coli* but Not for Catalysis. *Journal of Microbiology and Biotechnology*, 26(10), pp.1701–1707.

Newie, J. et al., 2017. Lipoxygenase 2 from *Cyanothece* sp. controls dioxygen insertion by steric shielding and substrate fixation. *Scientific Reports*, 7(1), p.2069.

Nielsen, H., 2017. Predicting Secretory Proteins with SignalP. In pp. 59–73.

Ohara, T. & Itoh, K., 2003. Significance of *Pseudomonas aeruginosa* colonization of the gastrointestinal tract. *Internal Medicine (Tokyo, Japan)*, 42(11), pp.1072–6.

Oliver, J.D. & Colwell, R.R., 1973. Extractable lipids of gram-negative marine bacteria: Fatty-acid composition. *International Journal of Systematic Bacteriology*, 23(4), pp.442–458.

Pekárová, M. et al., 2015. Mutagenesis of triad determinants of rat Alox15 alters the specificity of fatty acid and phospholipid oxygenation. *Archives of Biochemistry and Biophysics*, 571, pp.50–57.

Pistorius, E.K. & Axelrod, B. (1974). Iron, an essential component of lipoxygenase., 249(1973), pp. 3183–3186.

Porta, H. & Rocha-Sosa, M. (2001). Lipoxygenase in bacteria: a horizontal transfer event? *Microbiology* (Reading, England), 147, pp. 3199–3200.

Porta, H. & Rocha-Sosa, M. (2002). Plant lipoxygenases. Physiological and molecular features. *Plant Physiology*, 130(1), pp. 15–21.

Raychowdhury, M.K., Goswami, R. & Chakrabarti, P. (1985). Effect of unsaturated fatty acids in growth inhibition of some penicillin-resistant and sensitive bacteria. *The Journal of Applied Bacteriology*, 59(2), pp. 183–8.

Rejman, J. et al. (2007). *Pseudomonas aeruginosa* Infection Destroys the Barrier Function of Lung Epithelium and Enhances Polyplex-Mediated Transfection. *Human Gene Therapy*, 18(7), pp. 642–652.

Romano, M. (2010). Lipoxin and aspirin-triggered lipoxins. *The Scientific World Journal*, 10, pp. 1048–1064.

Schewe, T. et al. (1975). A lipoxygenase in rabbit reticulocytes which attacks phospholipids and intact mitochondria. *FEBS Letters*, 60(1), pp. 149–152.

Serhan, C.N. et al. (2012). Macrophage proresolving mediator maresin 1 stimulates tissue regeneration and controls pain. *FASEB Journal : official publication of the Federation of American Societies for Experimental Biology*, 26(4), pp. 1755–1765.

Sharma, G. et al. (2014). *Pseudomonas aeruginosa* biofilm: potential therapeutic targets. *Biologicals: Journal of the International Association of Biological Standardization*, 42(1), pp. 1–7.

Shimahara, K. (1964). Peroxidation of soy bean oil by lipoxygenase-forming bacteria (Gram-negative, rod-shaped) isolated from garbage. *Kogio Kagaku Zasshi*, 67, pp. 1164–1168.

Shimahara, K. & Hashizume, Y. (1973). Properties of a lipoxygenase-like enzyme produced by *Pseudomonas aeruginosa* strain A-4. *Journal of Fermentation Technology*, 51, pp. 183–189.

Sloane, D.L. et al. (1991). A primary determinant for lipoxygenase positional specificity. *Nature*, 354(6349), pp. 149–152.

Sloane, D.L. et al. (1995). Conversion of human 15-lipoxygenase to an efficient 12-lipoxygenase: the side-chain geometry of amino acids 417 and 418 determine positional specificity. *Protein Engineering*, 8(3), pp. 275–282.

Su, C. & Oliw, E.H. (1998). Manganese lipoxygenase. Purification and characterization. *The Journal of Biological Chemistry*, 273(21), pp. 13072–13079.

Suzuki, H. et al. (1994). Site-directed mutagenesis studies on the iron-binding domain and the determinant for the substrate oxygenation site of porcine leukocyte arachidonate 12-lipoxygenase. *Biochimica et Biophysica Acta*, 1210(3), pp. 308–316.

Takahashi, Y. et al. (1993). Investigation of the oxygenation of phospholipids by the porcine leukocyte and human platelet arachidonate 12-lipoxygenases. *European Journal of Biochemistry/ FEBS*, 218(1), pp. 165–171.

Vance, R.E. et al. (2004). The opportunistic pathogen *Pseudomonas aeruginosa* carries a secretable arachidonate 15-lipoxygenase. *Proceedings of the National Academy of Sciences of the United States of America*, 101(7), pp. 2135–2139.

Veesenmeyer, J.L. et al. (2009). *Pseudomonas aeruginosa* virulence and therapy: evolving translational strategies. *Critical Care Medicine*, 37(5), pp. 1777–86.

Vidal-Mas, J. et al. (2005). Cloning and expression of a lipoxygenase from *Pseudomonas aeruginosa* 42A2. *Antonie van Leeuwenhoek*, 87(3), pp. 245–251.

Vogel, R. et al. (2010). Applicability of the triad concept for the positional specificity of mammalian lipoxygenases. *The Journal of Biological Chemistry*, 285(8), pp. 5369–5376.

Volkov, A. et al. (2010). Myosin cross-reactive antigen of *Streptococcus pyogenes* M49 encodes a fatty acid double bond hydratase that plays a role in oleic acid detoxification and bacterial virulence. *The Journal of Biological Chemistry*, 285(14), pp. 10353–61.

Wasternack, C. & Feussner, I. (2018). The Oxylipin Pathways: Biochemistry and Function. *Annual Review of Plant Biology*, 69, pp. 363–386.

Wecksler, A.T. et al. (2008). Substrate specificity changes for human reticulocyte and epithelial 15-lipoxygenases reveal allosteric product regulation. *Biochemistry*, 47(28), pp. 7364–7375.

Yamamoto, S., Juranek, I. & Suzuki, H. (2002). Affinities of mammalian arachidonate oxygenases for molecular oxygen. *International Congress Series*, 1233, pp. 337–342.

Yoo, S., Lim, J.Y. & Hwang, S.W. (2013). Resolvins: Endogenously-Generated Potent Painkilling Substances and their Therapeutic Perspectives. *Current Neuropharmacology*, 11(6), pp. 664–676.

Zheng, Y. et al. (2008). A 49-kDa mini-lipoxygenase from *Anabaena* sp. PCC 7120 retains catalytically complete functionality. *Journal of Biological Chemistry*, 410(8), pp. 347–357.

ABOUT THE EDITORS

Chee Fah Wong, PhD

Associate Professor, Department of Biology, Faculty of Science and Mathematics, Universiti Pendidikan Sultan Idris, Perak, Malaysia

Chee Fah Wong is an Associate Professor in Protein Chemistry, Universiti Pendidikan Sultan Idris, Malaysia. He was graduated from Universiti Putra Malaysia in Microbial Biotechnology and completed his post-doctoral studies in Kyoto University, Japan, in the field of Structural Biology. His efforts have been targeted on the search of extremophilic enzymes, which are superior to the traditional catalysts because they can perform industrial processes even under harsh conditions, under which conventional proteins are completely denatured. This may be particularly true for several enzymes, the thermophilic aminoacylase and organic solvent tolerant proteases and lipases from *Pseudomonas*, where its ability to remain catalytically active under extremes of temperature, salinity, pH and organic solvent is seen as one potential answer to the biotechnological conflict between industrial conditions and the fragility of biological components. Using these enzymes as models, the mechanisms of protein structural stability and adaptation in extreme conditions can be explored and unveiled. He also develops passion into electrophysiology and physical interaction of neurotransmitter release-related proteins/complexes and drug drug for antiobesity treatment.

Hamidah Idris, PhD

Senior Lecturer, Department of Biology, Faculty of Science and Mathematics, Universiti Pendidikan Sultan Idris, Perak, Malaysia

Hamidah Idris is a Senior Lecturer at Universiti Pendidikan Sultan Idris, Malaysia in Microbiology field. She started her tertiary education at National University of Malaysia where she graduated with Bachelor of Science with Honours (Microbiology) and Master of Science (Microbiology). With the strong interest in bacterial taxonomy, biodiversity and natural products, she attended Newcastle University, UK, where she received her Ph.D in Microbiology in 2016. Throughout her newly build career, she has publications in areas including actinobacterial species description, metagenomics studies and microbial natural products. Experience in doing research that integrate microbiology experiment with different fields including bioinformatics, chemistry, environmental science and statistic allow her to explore more about the unseen microbial world. Research involving bacteria especially from extreme environments has been her passion as there are a lot of questions yet to be answered, including the type of bacteria present in the unravelled environment, role in ecosystem, influencing factors and ability to produce useful substances.

INDEX

#

A

C

D

H

I

K

L

M

N

O

P

Q

R

T

U

V

W

Y

Z

β

Related Nova Publications

Staphylococcus aureus: Infections, Treatment and Risk Assessment

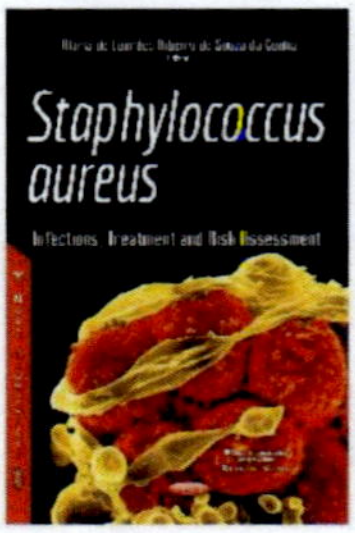

Editor: Maria de Lourdes Ribeiro de Souza da Cunha

Series: Bacteriology Research Developments

Book Description: This book provides an overview of the different topics of staphylococcal research in recent years. Severe infections caused by methicillin-resistant Staphylococcus aureus (MRSA) have become a global health problem in the twenty-first century.

Softcover ISBN: 978-1-63485-959-2
Retail Price: $95

Salmonella enterica: Molecular Characterization, Role in Infectious Diseases and Emerging Research

Editor: Faruk van Doleweerd

Series: Bacteriology Research Developments

Book Description: The bacterium Salmonella enterica is a serious bacterial pathogen and causative agent of salmonellosis infections on a worldwide scale.

Softcover ISBN: 978-1-53613-084-3
Retail Price: $82